BARRON'S

Painless
Algebra

Fifth Edition

Lynette Long, Ph.D.

Published by Kaplan, Inc., d/b/a Barron's Educational Series
1515 W. Cypress Creek Road
Fort Lauderdale, FL 33309
www.barronseduc.com

ISBN: 978-1-5062-6806-4

10 9 8 7 6 5 4 3 2

Kaplan, Inc., d/b/a Barron's Educational Series print books are available at special
quantity discounts to use for sales promotions, employee premiums, or educational
purposes. For more information or to purchase books, please call the Simon &
Schuster special sales department at 866-506-1949.

Contents

How to Use This Book

Painless algebra? Impossible, you think. Not really. I have been teaching math or teaching teachers how to teach math for more than twenty years. Math is easy . . . or at least it can be with the help of this book!

Whether you are learning algebra for the first time, or you are trying to remember what you've learned but have forgotten, this book is for you. It provides a clear introduction to algebra that is both fun and instructive. Don't be afraid. Dive in—it's painless!

Painless Icons and Features

This book is designed with several unique features to help make learning algebra easy.

 **PAINLESS TIP**

You will see Painless Tips throughout the book. These include helpful tips, hints, and strategies on the surrounding topics.

CAUTION—Major Mistake Territory!

Caution boxes will help you avoid common pitfalls or mistakes. Be sure to read them carefully.

1+2=3 MATH TALK!

These boxes translate "math talk" into plain English to make it even easier to understand math.

 **REMINDER**

Reminders will call out information that is important to remember. Each reminder will relate to the current chapter or will reference key information you learned in a previous chapter.

 **BRAIN TICKLERS**

There are brain ticklers throughout each chapter in the book. These quizzes are designed to make sure you understand what you've just learned and to test your progress as you move forward in the chapter. Complete all the Brain Ticklers and check your answers. If you get any wrong, make sure to go back and review the topics associated with the questions you missed.

PAINLESS STEPS

Complex procedures are divided into a series of painless steps. These steps help you solve problems in a systematic way. Follow the steps carefully, and you'll be able to solve most algebra problems.

EXAMPLES

Most topics include examples with solutions. If you are having trouble, research shows that writing or copying the problem may help you understand it.

ILLUSTRATIONS

Painless Algebra is full of illustrations to help you better understand algebra topics. You'll find graphs, charts, and more instructive illustrations to help you along the way.

SIDEBARS

These shaded boxes contain extra information that relates to the surrounding topics. Sidebars can include detailed examples or practice tips to help keep algebra interesting and painless.

Chapter Breakdown

Chapter One is titled "A Painless Beginning," and it really is. It is an introduction to numbers and number systems. It will teach you how to perform simple operations on both numbers and variables painlessly, and by the end of the chapter you will know what "Please Excuse My Dear Aunt Sally" means.

Chapter Two shows you how to add, subtract, multiply, and divide both positive and negative numbers. The only trick is remembering which sign the answer has, and with a little practice you'll be a whiz.

Chapter Three teaches you how to solve equations. Think of an equation as a number sentence that contains a mystery number. All you have to do is figure out the value of the mystery number. Just follow a few simple, painless steps to success.

Chapter Four shows you how to solve inequalities. What happens when the mystery number is not part of an equation but instead is part of a number sentence in which one part of the sentence is greater than the other part? Now what could the mystery number be? Does the solution sound complicated? Trust me. It's painless.

Chapter Five is all about graphing. You will learn what coordinate axes are, and learn how to graph horizontal, vertical, and diagonal lines. You'll even learn how to graph inequalities, so get out a pencil and a ruler and get started.

Chapter Six shows you how to solve systems of equations and in-equalities. Systems of equations are two or more equations taken together. You try to find a single answer that will make them all true. You'll learn to solve systems in many different ways. It's fun, because no matter how you solve them you always get the same answer. That's one of the magical things about mathematics!

Chapter Seven deals with exponents. What happens when you multiply a number by itself seven times? You can write 2 times 2 times 2 times 2 times 2 times 2 times 2, or you can use an exponent and write 2^7, which is two to the seventh power. This chapter will introduce you to exponents, which are amazing shortcuts, and teach you how to work with them.

Chapter Eight is about roots and radicals. In mathematics, roots are not tree roots, but the opposite of exponents. What are radicals? You'll have to wait until Chapter Eight to find out.

Chapter Nine shows you how to solve quadratic equations. *Quadratic* is a big word, but don't get nervous. It's just a fancy name mathematicians give to an equation with an *x*-squared term in it.

Chapter Ten discusses advanced topics in algebra. If you're interested in what number comes next in a series of numbers, a magic triangle discovered by Pascal, or how to solve a matrix or a function, check out this chapter.

Chapter 1

A Painless Beginning

Algebra is a language. In many ways, mastering algebra is just like learning French, Italian, or German, or maybe even Japanese. To understand algebra, you need to learn how to read it and how to change Plain English into Math Talk and Math Talk back into Plain English.

In algebra, a letter is often used to stand for a number. The letter used to stand for a number is called a *variable*. You can use any letter, but a, b, c, n, x, y, and z are the most commonly used letters. In the following sentences, different letters are used to stand for numbers.

$x + 3 = 5$ x is a variable.

$a - 2 = 6$ a is a variable.

$y \div 3 = 4$ y is a variable.

$x + y = 7$ Both x and y are variables.

When you use a letter to stand for a number, you don't know what number the letter represents. Think of the letters x, a, y, and z as mystery numbers.

A variable can be part of an expression, an equation, or an inequality. A mathematical expression is part of a mathematical sentence, just as a phrase is part of an English sentence. Here are a few examples of mathematical expressions: $3x$, $x + 5$, $x - 2$, $x \div 10$. In each of these expressions it is impossible to know what x is. The variable x could be any number.

Mathematical expressions are named based on how many terms they have. A *monomial expression* has one term.

> 3 is a monomial expression.
>
> $6x$ is a monomial expression.

A *binomial expression* has two unlike terms combined by an addition or subtraction sign.

> $x + 3$ is a binomial expression.
>
> $a - 4$ is a binomial expression.
>
> $x + y$ is a binomial expression.

A *trinomial expression* has three unlike terms combined by addition and/or subtraction signs.

> $x + y - 3$ is a trinomial expression.
>
> $2x - 3y + 7$ is a trinomial expression.
>
> $4a - 5b + 6c$ is a trinomial expression.

A *polynomial expression* has one, two, or more unlike terms combined by addition and/or subtraction signs. Monomials, binomials, and trinomials are polynomial expressions. The following are also polynomial expressions:

> $x + y + z - 4$ is a polynomial expression.
>
> $2a + 3b - 4c + 2$ is a polynomial expression.

A *mathematical sentence* contains two mathematical phrases joined by an equals sign or an inequality sign. An *equation* is a mathematical sentence in which the two phrases are joined by an equals sign. Notice that the word *equation* starts the same way as the word *equal*.

> $3 + 6 = 9$ is an equation.
>
> $x + 1 = 2$ is an equation.
>
> $7x = 14$ is an equation.
>
> $0 = 0$ is an equation.
>
> $4x + 3$ is not an equation. It does not have an equals sign. It is a mathematical expression.

Some equations are true and some equations are false.

$3 = 2 + 1$ is a true equation.

$3 + 5 = 7$ is an equation, but it is false.

$x + 1 = 5$ is an equation. It could be true or it could be false.

Whether $x + 1 = 5$ is true or false depends on the value of x. If x is 4, the equation $x + 1 = 5$ is true. If $x = 0$, $x + 1 = 5$ is false.

An *inequality* is a mathematical sentence in which two phrases are joined by an inequality symbol. The inequality symbols are *greater than*, ">"; *greater than or equal to*, "≥"; *less than*, "<"; and *less than or equal to*, "≤."

Six is greater than five is written as $6 > 5$. Seven is less than ten is written as $7 < 10$.

Mathematical Operations

In mathematics, there are four basic operations: addition, subtraction, multiplication, and division. When you first learned to add, subtract, multiply, and divide, you used the symbols $+$, $-$, $\times$, and $\div$. In algebra, addition is still indicated by the plus ($+$) sign and subtraction is still indicated by the minus ($-$) sign.

Addition

When you add you can add only *like terms*.

Terms that consist only of numbers are like terms.

$5, 3, 0.4$, and $\frac{1}{2}$ are like terms.

Terms that use the same variable to the same degree (with the same exponents) are like terms.

$3z$, $-6z$, and $\frac{1}{2}z$ are like terms.

Terms with different exponents are unlike terms.

x^2 and x^3 and x^{-1} are unlike terms.

A number and a variable are unlike terms.

> 7 and x are unlike terms.

Terms that use different variables are unlike terms.

> $3z$, b, and $-2x$ are unlike terms.

You can add any numbers.

> $3 + 6 = 9$

> $5 + 2 + 7 + 6 = 20$

You can also add variables as long as they are the same variable. To add like variables, just add the *coefficients*. The coefficient is the number in front of the variable.

> In the expression $7a$, 7 is the coefficient and a is the variable.
>
> In the expression $\frac{1}{2}y$, $\frac{1}{2}$ is the coefficient and y is the variable.
>
> In the expression x, 1 is the coefficient and x is the variable.

Now note how like terms are added to simplify the following expressions.

> $3x + 7x = 10x$
>
> $4x + 12x + \frac{1}{2}x = 16\frac{1}{2}x = \frac{33}{2}x$

You cannot simplify $3x + 5y$ because the variables are not the same.

You cannot simplify $3x + 4$ because $3x$ is a variable expression and 4 is a number.

CAUTION—Major Mistake Territory!

When adding expressions with the same variable, just add the coefficients and attach the variable to the new coefficient. Do not put two variables at the end.

$$2x + 3x \neq 5xx$$
$$2x + 3x = 5x$$

Note: An equals sign with a slash through it ($\neq$) means "not equal to."

Subtraction

You can also subtract like terms.

You can subtract one number from another number.

$$7 - 3 = 4$$
$$12 - 12 = 0$$

You can subtract one variable expression from another variable expression. Just subtract the coefficients and keep the variable the same.

$$7a - 4a = 3a$$
$$3x - x = 2x \text{ (Remember: the coefficient of } x \text{ is 1.)}$$
$$4y - 4y = 0y = 0$$

You cannot simplify $3x - 4y$ because the terms do not have the same variable. You cannot simplify $100 - 7b$ because 100 and $7b$ are not like terms.

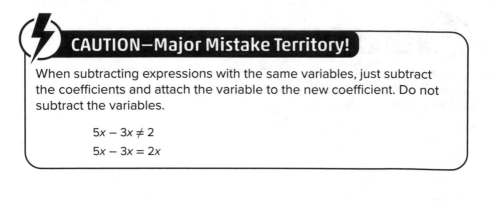

CAUTION—Major Mistake Territory!

When subtracting expressions with the same variables, just subtract the coefficients and attach the variable to the new coefficient. Do not subtract the variables.

$$5x - 3x \neq 2$$
$$5x - 3x = 2x$$

BRAIN TICKLERS   Set # 1

Add or subtract each of these algebraic expressions.

1. $3x + 7x$
2. $4x + x$
3. $3x - 3x$

4. $10x - x$
5. $3x + 2$

(Answers are on page 25.)

Multiplication

An $\times$ is seldom used to indicate multiplication. It is too easy to confuse $\times$, which means "multiply," with x the variable. To avoid this problem, mathematicians use other ways to indicate multiplication. Here are three ways to write "multiply."

1. A centered dot ($\cdot$) indicates "multiply."

    $$3 \cdot 5 = 15$$

2. Writing two letters or a letter and a number next to each other is another way of saying "multiply."

    $$7b = 7 \cdot b$$

3. Writing a letter or a number before a set of parentheses says "multiply."

    $$6(2) = 12$$

You can multiply like and unlike terms.

You can multiply any two numbers.

$$3(4) = 12$$

$$8\left(\frac{1}{2}\right) = 4$$

You can multiply any two variables.

$$x \cdot x = (x)(x) = x^2$$

$$a \cdot b = (a)(b) = ab$$

You can multiply a number and a variable.

$$3 \cdot x = 3x$$

You can even multiply two expressions if one is a number and the other is a variable with a coefficient—for example, 4 times $3x$. To multiply these expressions requires two *painless* steps.

1. Multiply the coefficients.

2. Attach the variable at the end of the answer.

Watch:

3 times $5x$

First multiply the coefficients.
$$3 \cdot 5 = 15$$

Next attach the variable at the end of the answer.
$$15x$$

Solution: $3 \cdot 5x = 15x$

$6y$ times 2

First multiply the coefficients.
$$6 \cdot 2 = 12$$

Next attach the variable at the end of the answer.
$$(12)\,(y) = 12y$$

Solution: $6y \cdot 2 = 12y$

You can also multiply two expressions even if they both have numbers and variables. To multiply these expressions takes three painless steps.

Step 1: Multiply the coefficients.

Step 2: Multiply the variables.

Step 3: Combine the two answers.

Learn:

Multiply 3x times 2y.

First multiply the coefficients.
$$3 \cdot 2 = 6$$

Next multiply the variables.
$$x \cdot y = xy$$

Combine the two answers by multiplying them.
$$6xy$$

Solution: $3x \cdot 2y = 6xy$

Multiply 4x times 5x.

First multiply the coefficients.
$$4 \cdot 5 = 20$$

Next multiply the variables.
$$x(x) = x^2$$

Combine the two answers by multiplying them.
$$20x^2$$

Solution: $4x \cdot 5x = 20x^2$

Multiply 6x times y.

First multiply the coefficients, 6 and 1.
$$6 \cdot 1 = 6$$

Next multiply the variables.
$$x \cdot y = xy$$

Combine the two answers by multiplying them.
$$6xy$$

Solution: $6x \cdot y = 6xy$

Division

The division sign, ÷, means "divide." The expression 6 ÷ 6 is read as "six divided by six." In algebra, ÷ is seldom used to indicate division. Instead, a slash mark, /, or a horizontal fraction bar, —, is used.

$6/6$ or $\frac{6}{6}$ means "six divided by six."

$a/3$ or $\frac{a}{3}$ means "a divided by three."

You can divide like and unlike terms in algebra.

You can divide any two numbers.

$$3 \text{ divided by } 4 = \frac{3}{4}$$

$$12 \text{ divided by } 6 = \frac{12}{6} = 2$$

You can divide any two of the same variables.

$$x \text{ divided by } x = \frac{x}{x} = 1$$

You can also divide any two different variables.

$$a \text{ divided by } b = \frac{a}{b}$$

You can also divide two variables with coefficients. To divide these expressions requires three steps.

Step 1: Divide the coefficients.

Step 2: Divide the variables.

Step 3: Multiply the answers.

Watch. The division is *painless*.

Divide $3x$ by $4x$.

First divide the coefficients.

$\quad$ 3 divided by 4 $= \dfrac{3}{4}$

Next divide the variables.

$\quad\quad x$ divided by $x = \dfrac{x}{x} = 1$

Finally, multiply the two answers.

$\quad\quad \dfrac{3}{4}$ times $1 = \dfrac{3}{4}$

$\quad\quad$ Solution: $\dfrac{3x}{4x} = \dfrac{3}{4}$

Divide $8x$ by $2y$.

First divide the coefficients.

$\quad\quad$ 8 divided by 2 $= \dfrac{8}{2} = 4$ or $\dfrac{4}{1}$

Next divide the variables.

$\quad\quad x$ divided by $y = \dfrac{x}{y}$

Multiply the two answers.

$\quad\quad \left(\dfrac{4}{1}\right)\left(\dfrac{x}{y}\right) = \dfrac{4x}{y}$

$\quad\quad$ Solution: $\dfrac{8x}{2y} = \dfrac{4x}{y}$

Divide $12xy$ by x.

Remember, if there is no coefficient in front of a variable, it is one!

First divide the coefficients, 12 and 1.

$\quad\quad$ 12 divided by 1 $= \dfrac{12}{1} = 12$

Next divide the variables.

$\quad\quad xy$ divided by $x = \dfrac{xy}{x} = y$, since $\dfrac{x}{x} = 1$

Finally, multiply the two answers.

$\quad\quad (12)(y) = 12y$

$\quad\quad$ Solution: $\dfrac{12xy}{x} = 12y$

BRAIN TICKLERS   Set # 2

Solve these multiplication and division problems.

1. $3x$ times $4y$

2. $6x$ times $2x$

3. $2x$ times 5

4. $7x$ divided by $7x$

5. $4xy$ divided by $2x$

6. $3x$ divided by 3

7. $8xy$ divided by y

(Answers are on page 25.)

Zero

Zero is an unusual number. It is neither positive nor negative. There are some rules about zero you should know. If zero is added to any number, the answer is that number. If zero is added to any variable, the answer is that variable.

$$5 + 0 = 5$$
$$x + 0 = x$$

If any number or variable is added to zero, the answer is that number or variable.

$$0 + 9 = 9$$
$$0 + \frac{1}{2} = \frac{1}{2}$$
$$0 + a = a$$

If zero is subtracted from any number or variable, the answer is that number or variable.

$$3 - 0 = 3$$
$$b - 0 = b$$

If a number or variable is subtracted from zero, the answer is the opposite of that number or variable.

$$0 - 3 = -3$$
$$0 - (-4) = 4$$
$$0 - b = -b$$

If any number or variable is multiplied by zero, the answer is always zero.

$$1,000(0) = 0$$
$$d(0) = 0$$
$$7xy \cdot 0 = 0$$

If zero is multiplied by any number or variable, the answer is always zero.

$$0 \cdot 7 = 0$$
$$0(x) = 0$$

If zero is divided by any number or variable, the answer is always zero.

$$0 \div 3 = 0$$
$$0 \div (-5) = 0$$
$$\frac{0}{f} = 0$$

If any number or variable is divided by zero, the result is undefined. You cannot divide by zero.

$$3 \div 0 = ?$$ Division by zero is undefined.

$$\frac{a}{0} = ?$$ Division by zero is undefined.

CAUTION—Major Mistake Territory!

You can never divide by zero. Division by zero is undefined. Don't get trapped into thinking $5 \div 0$ is 0 or $5 \div 0$ is 5; $5 \div 0$ is undefined.

BRAIN TICKLERS   Set # 3

All of these problems involve zero. Solve them.

1. $0 + a$

2. $a(0)$

3. $0 - a$

4. $\dfrac{0}{a}$

5. $(0)a$

6. $a - 0$

7. $\dfrac{a}{0}$

8. $a + 0$

(Answers are on page 25.)

Order of Operations

When you solve or simplify a mathematical sentence or expression, it's important that you do things in the correct order. The order in which you solve a problem may affect the answer.

Look at the following problem:

$$3 + 1 \cdot 6$$

You read the problem as "three plus one times six," but does it mean "the quantity three plus one times six," which is written as $(3 + 1) \times 6$, or "three plus the quantity one times six," which is written as $3 + (1 \times 6)$?

These two problems have different answers.

$$(3 + 1) \times 6 = 24$$
$$3 + (1 \times 6) = 9$$

Which is the correct answer?

Mathematicians have agreed on a certain sequence, called the *Order of Operations*, to be used in solving mathematical problems. Without the Order of Operations, several different answers would be possible when computing mathematical expressions. The Order of Operations tells you how to simplify any mathematical expression in four easy steps.

Step 1: Do everything in parentheses.

In the problem $7(6 - 1)$, subtract first, then multiply.

Step 2: Compute the value of any exponential expressions.

In the problem $5 \cdot 3^2$, square the three first and then multiply by five.

Step 3: Multiply and/or divide. Start at the left and go to the right.

In the problem $5 \cdot 2 - 4 \cdot 3$, multiply five times two and then multiply four times three. Subtract last.

Step 4: Add and/or subtract. Start on the left and go to the right.

In the problem $6 - 2 + 3 - 4$, start with six, subtract two, add three, and subtract four.

To remember the Order of Operations just remember the sentence "**P**lease **E**xcuse **M**y **D**ear **A**unt **S**ally!" The first letter of each word tells you what to do next. The "P" in Please stands for parentheses. The "E" in Excuse stands for exponents. The "M" in My stands for multiply. The "D" in Dear stands for divide. The "A" in Aunt stands for add. The "S" in Sally stands for subtract.

If you remember your Aunt Sally, you'll never forget the Order of Operations. There is one trick to keep in mind: multiply and divide at the same time, and add and subtract at the same time.

1+2=3 MATH TALK!

Watch as these mathematical expressions are changed from Math Talk into Plain English.

$3(5 + 2) - 7$

three times the quantity five plus two, then that quantity minus seven

$3 - 7(7) - 2$

three minus the quantity seven times seven, then that quantity minus two

$6 - 3^2 + 2$

six minus the quantity three squared, then that quantity plus two

Watch how the following expression is evaluated.

3(5 − 2) + 6 · 1

Step 1: Do what is inside the parentheses.

$$5 - 2 = 3$$
$$3(5 - 2) + 6 \cdot 1 = 3(3) + 6 \cdot 1$$

Step 2: Compute the values of any exponential expressions. There are no exponential expressions.

Step 3: Multiply and/or divide from left to right.

$$3(3) + 6 \cdot 1 = 9 + 6$$

Step 4: Add and/or subtract from left to right.

$$9 + 6 = 15$$
Solution: $3(5 - 2) + 6 \cdot 1 = 15$

Watch how the following expression is evaluated.

(4 − 1)² − 2 · 3

Step 1: Do what is inside the parentheses.

$$(4 - 1)^2 - 2 \cdot 3 = 3^2 - 2 \cdot 3$$

Step 2: Compute the value of any exponential expressions.

$$3^2 - 2 \cdot 3 = 9 - 2 \cdot 3$$

Step 3: Multiply and/or divide from left to right.

$$9 - 2 \cdot 3 = 9 - 6$$

Step 4: Add and/or subtract from left to right.

$$9 - 6 = 3$$
Solution: $(4 - 1)^2 - 2 \cdot 3 = 3$

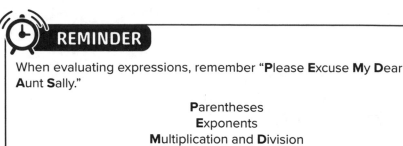

REMINDER

When evaluating expressions, remember "**P**lease **E**xcuse **M**y **D**ear **A**unt **S**ally."

Parentheses
Exponents
Multiplication and **D**ivision
Addition and **S**ubtraction

BRAIN TICKLERS   Set # 4

Compute the values of the following expressions. Keep in mind the Order of Operations. Remember: "Please Excuse My Dear Aunt Sally."

1. $4 \div (2 + 2) - 1$
2. $3 + 12 - 5 \cdot 2$
3. $16 - 2 \cdot 4 + 3$
4. $6 + 5^2 - 12 + 4$
5. $(4 - 3)^2(2) - 1$

(Answers are on page 25.)

Properties of Numbers

Five properties of numbers are important in the study of algebra.

The Commutative Property of Addition
The Commutative Property of Multiplication
The Associative Property of Addition
The Associative Property of Multiplication
The Distributive Property of Multiplication over Addition

What does each of these properties say?

The commutative property of addition

The Commutative Property of Addition states that, no matter in what order you add two numbers, the sum is always the same. In other words, three plus four is equal to four plus three. Six plus two is equal to two plus six.

Examples:

$$3 + 5 = 5 + 3 \text{ because } 8 = 8.$$

$$5x + 3 = 3 + 5x$$

> **1+2=3 MATH TALK!**
>
> The Commutative Property of Addition states that, given any two numbers a and b, a plus b is equal to b plus a. In plain English, the order you add two numbers together doesn't matter.

> **PAINLESS TIP**
>
> Subtraction is not commutative. The order of the numbers in a subtraction problem *does* make a difference.
>
> $6 - 3$ is not the same as $3 - 6$.
>
> $5 - 0$ is not the same as $0 - 5$.

The commutative property of multiplication

The Commutative Property of Multiplication states that, no matter in what order you multiply two numbers, the answer is always the same.

Examples:

$$3 \cdot 5 = 5 \cdot 3$$

$$6(y) = y(6)$$

$$\frac{1}{2}(4) = 4\left(\frac{1}{2}\right)$$

1+2=3 **MATH TALK!**

The Commutative Property of Multiplication states that $a(b) = b(a)$. Given any two numbers, a and b, a times b is equal to b times a. In plain English, the order you multiply two numbers together doesn't matter.

PAINLESS TIP

Division is not commutative. The order of the numbers in a division problem *does* make a difference.

$5 \div 10$ is not the same as $10 \div 5$.

$\frac{6}{2}$ is not the same as $\frac{2}{6}$.

$\frac{a}{2}$ is not the same as $\frac{2}{a}$.

The associative property of addition

The Associative Property of Addition states that, when you add three numbers, no matter how you group them, the answer is still the same.

Examples:

$$(3 + 5) + 2 = 3 + (5 + 2)$$
$$(1 + 8) + 4 = 1 + (8 + 4)$$

1+2=3 **MATH TALK!**

The Associative Property of Addition states that $(a + b) + c = a + (b + c)$. If you first add a and b and then add c to the total, the answer is the same as if you first add b and c and then add the total to a. In plain English, if you add a group of numbers together, group them however you want. The answer will be the same.

The associative property of multiplication

The Associative Property of Multiplication states that, when you multiply three numbers, no matter how you group them, the product is always the same.

Examples:

$$(3 \cdot 2)6 = 3(2 \cdot 6)$$

$$(5 \cdot 4)2 = 5(4 \cdot 2)$$

> **1+2=3 MATH TALK!**
>
> The Associative Property of Multiplication states that $(a \cdot b)c = a(b \cdot c)$. If you first multiply a and b and then multiply the product by c, the answer is the same as if you first multiply b and c and then multiply a by the product. In plain English, when you multiply several numbers together, it doesn't matter how you group them.

The distributive property of multiplication over addition

The Distributive Property of Multiplication over Addition states that, when you multiply a single number such as 3 or a monomial such as $5x$ by a binomial expression such as $(2 + x)$, the answer is the monomial or number (3) times the first term of the binomial expression (2) plus the monomial (3) times the second term of the binomial expression (x). Just remember to multiply the number or expression outside the parentheses by each of the numbers or expressions inside the parentheses.

Examples:

$$3(5 + 2) = 3 \cdot 5 + 3 \cdot 2 = 21$$

$$\frac{1}{2}(4 + 1) = \frac{1}{2}(4) + \frac{1}{2}(1) = \frac{5}{2}$$

$$4(6 - 2) = 4(6) + 4(-2) = 24 - 8 = 16$$

$$6(3 + x) = 6(3) + 6(x) = 18 + 6x$$

$$5y(2x + 3) = 5y(2x) + 5y(3) = 10yx + 15y$$

1+2=3 **MATH TALK!**

The Distributive Property of Multiplication over Addition states that
$a(b + c) = ab + ac$. Multiplying a by the quantity b plus c is equal to
a times b plus a times c.

BRAIN TICKLERS   Set # 5

Use the following abbreviations as instructed below.

CA = The Commutative Property of Addition

CM = The Commutative Property of Multiplication

AA = The Associative Property of Addition

AM = The Associative Property of Multiplication

DM/A = The Distributive Property of Multiplication over Addition

Next to each mathematical equation, write the abbreviation for the property
the equation represents. Be careful—some of the problems are tricky.

_____ 1. $6(5 + 1) = 6(5) + 6(1)$

_____ 2. $4 + (3 + 2) = (4 + 3) + 2$

_____ 3. $5 + 3 = 3 + 5$

_____ 4. $3(5 \cdot 1) = (3 \cdot 5)1$

_____ 5. $7(3) = 3(7)$

_____ 6. $6(4 + 3) = 6(3 + 4)$

(Answers are on page 26.)

Number Systems

There are six different number systems.

The natural numbers
The whole numbers
The integers
The rational numbers
The irrational numbers
The real numbers

The natural numbers

The natural numbers are $1, 2, 3, 4, 5, \ldots$.
The three dots, $\ldots$, mean "continue counting forever."
The natural numbers are sometimes called the *counting numbers*
  because they are the numbers that you use to count.

Examples:

7 and 9 are natural numbers.

$0, \frac{1}{2}$, and -3 are *not* natural numbers.

The whole numbers

The whole numbers are $0, 1, 2, 3, 4, 5, 6, \ldots$.
The whole numbers are the natural numbers plus zero.
All of the natural numbers are whole numbers.

Examples:

$0, 5, 23$, and $1,001$ are whole numbers.

$-4, \frac{1}{3}$, and 0.2 are *not* whole numbers.

The integers

The integers are the natural numbers, their opposites, and zero.
The integers are $\ldots, -3, -2, -1, 0, 1, 2, 3, \ldots$.
All of the whole numbers are integers.
All of the natural numbers are integers.

Examples:

$-62, -12, 27$, and 83 are integers.

$-\frac{3}{4}, \frac{1}{2}$, and $\sqrt{2}$ are *not* integers.

The rational numbers

The rational numbers are any numbers that can be expressed
 as the ratios of two whole numbers.
All of the integers are rational numbers.
All of the whole numbers are rational numbers.
All of the natural numbers are rational numbers.

Examples:

3 can be written as $\frac{3}{1}$, so 3 is a rational number.

$-27, -12\frac{1}{2}, -\frac{1}{3}, \frac{1}{4}, 7, 4\frac{4}{5}$, and 1,000,000 are all rational
 numbers.

$\sqrt{2}$ and $\sqrt{3}$ are *not* rational numbers.

The irrational numbers

The irrational numbers are numbers that cannot be expressed as
 the ratios of two whole numbers.
The rational numbers are not irrational numbers.
The integers are not irrational numbers.
The whole numbers are not irrational numbers.
The natural numbers are not irrational numbers.

Examples:

$-\sqrt{2}, \sqrt{2}$, and $\sqrt{3}$ are irrational numbers.

$-41, -17\frac{1}{2}, -\frac{3}{8}, \frac{1}{5}, 4, \frac{41}{7}$, and 1,247 are *not* irrational numbers.

The real numbers

The real numbers are a combination of all the number systems.
The real numbers are the natural numbers, whole numbers,
 integers, rational numbers, and irrational numbers.
Every point on the number line is a real number.
All of the irrational numbers are real numbers.
All of the rational numbers are real numbers.
All of the integers are real numbers.
All of the whole numbers are real numbers.
All of the natural numbers are real numbers.

Examples:

$-53, -\dfrac{17}{3}, 4\dfrac{1}{2}, -\sqrt{2}, -\dfrac{3}{5}, 0, \dfrac{1}{6}, \sqrt{3}, 4, \dfrac{41}{7}$ and $1{,}247$ are all real numbers.

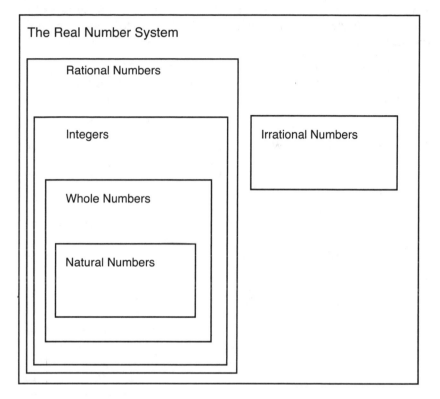

The Real Number System

Rational Numbers

Integers

Irrational Numbers

Whole Numbers

Natural Numbers

PAINLESS TIP

Is five a whole number or a natural number? Five is both a whole number and a natural number. A number can belong to more than one number system at the same time.

Is six a whole number or a rational number? Six is both a whole number and a rational number. Six can be written as 6 or as $\dfrac{6}{1}$.

BRAIN TICKLERS   Set # 6

Using the abbreviations below, circle the number system or systems to which each number belongs. Some numbers belong to more than one number system.

N = Natural numbers   Ra = Rational numbers
W = Whole numbers   Ir = Irrational numbers
In = Integers   Re = Real numbers

1.	3	N	W	In	Ra	Ir	Re
2.	−7	N	W	In	Ra	Ir	Re
3.	0	N	W	In	Ra	Ir	Re
4.	$-\frac{1}{4}$	N	W	In	Ra	Ir	Re
5.	$\frac{3}{8}$	N	W	In	Ra	Ir	Re
6.	$\frac{6}{1}$	N	W	In	Ra	Ir	Re
7.	−4	N	W	In	Ra	Ir	Re
8.	6	N	W	In	Ra	Ir	Re
9.	$\sqrt{3}$	N	W	In	Ra	Ir	Re
10.	$\frac{12}{3}$	N	W	In	Ra	Ir	Re

(Answers are on page 26.)

BRAIN TICKLERS—THE ANSWERS
Set # 1, page 6

1. $3x + 7x = 10x$
2. $4x + x = 5x$
3. $3x - 3x = 0$
4. $10x - x = 9x$
5. $3x + 2 = 3x + 2$

Set # 2, page 11

1. $3x$ times $4y = 3x(4y) = 12xy$

2. $6x$ times $2x = 6x(2x) = 12x^2$

3. $2x$ times $5 = 2x(5) = 10x$

4. $7x$ divided by $7x = \dfrac{7x}{7x} = 1$

5. $4xy$ divided by $2x = \dfrac{4xy}{2x} = 2y$

6. $3x$ divided by $3 = \dfrac{3x}{3} = x$

7. $8xy$ divided by $y = \dfrac{8xy}{y} = 8x$

Set # 3, page 13

1. $0 + a = a$
2. $a(0) = 0$
3. $0 - a = -a$
4. $\dfrac{0}{a} = 0$
5. $(0)a = 0$
6. $a - 0 = a$
7. $\dfrac{a}{0}$ is undefined.
8. a

Set # 4, page 16

1. $4 \div (2 + 2) - 1 = \dfrac{4}{4} - 1 = 1 - 1 = 0$

2. $3 + 12 - 5 \cdot 2 = 3 + 12 - 10 = 15 - 10 = 5$

3. $16 - 2 \cdot 4 + 3 = 16 - 8 + 3 = 11$

4. $6 + 5^2 - 12 + 4 = 6 + 25 - 12 + 4 = 23$

5. $(4 - 3)^2(2) - 1 = (1)^2(2) - 1 = 1(2) - 1 = 2 - 1 = 1$

Set # 5, page 20

1. DM/A
2. AA
3. CA
4. AM
5. CM
6. CA

Set # 6, page 24

1. N, W, In, Ra, Re
2. In, Ra, Re
3. W, In, Ra, Re
4. Ra, Re
5. Ra, Re
6. N, W, In, Ra, Re
7. In, Ra, Re
8. N, W, In, Ra, Re
9. Ir, Re
10. N, W, In, Ra, Re

Chapter 2

The Integers

When you first learned to count, you counted with positive numbers, the numbers greater than zero: 1, 2, 3, 4, 5. That's why these numbers are called the counting numbers. You probably first learned to count to 10, then to 100, and maybe even to 1,000. Finally, you learned that you could continue counting forever, because after every number there is a number that is one larger than the number before it.

Well, there are negative numbers, too. Negative numbers are used to express cold temperatures, money owed, feet below sea level, and lots of other things. You count from negative one to negative ten with these numbers: $-1, -2, -3, -4, -5, -6, -7, -8, -9, -10$. You could continue counting until you reached -100 or $-1,000$, or you could continue counting forever. After every negative number there is another negative number that is one less than the number before it.

When you place positive numbers, negative numbers, and zero together, you have what mathematicians call the *integers*.

What Are the Integers?

The integers are made up of three groups of numbers:

- the positive integers
- the negative integers
- zero

The positive integers are

1, 2, 3, 4, . . .

Sometimes the positive integers are written like this:

+1, +2, +3, +4, . . .

Here is a graph of the positive integers:

Notice that the positive integers are only the counting numbers. A number between any two counting numbers is not an integer. For example, $2\frac{1}{2}$, which is between 2 and 3, is not an integer.

1+2=3 MATH TALK!

If there is no sign in front of a number, assume that the number is positive.

4 can also be written as +4 or (+4).

+4 is read as "positive 4."

+4 is *not* read as "plus 4."

The negative integers are

−1, −2, −3, −4, −5, . . .

Sometimes they are written like this:

(−1), (−2), (−3), (−4), (−5), . . .

Here is a graph of the negative integers:

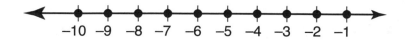

MATH TALK!

−3 can also be written as (−3).

−3 is read as "negative 3."

−3 is *not* read as "minus 3."

Zero is an integer, but it is neither positive nor negative.

Here is a graph of zero:

0

Here is a list representing all the integers:

$$\ldots, -4, -3, -2, -1, 0, 1, 2, 3, 4, \ldots$$

Here is a graph representing all the integers:

−4 −3 −2 −1 0 1 2 3 4

These numbers are integers:

$$0, 4, -7, -1{,}000, 365, \frac{10}{2}, -\frac{6}{3}, \frac{10}{10}, 123{,}456{,}789$$

These numbers are not integers:

$$7.2, \frac{6}{4}, -\frac{3}{8}, -1.2$$

CAUTION—Major Mistake Territory!

The number $\frac{6}{2}$ is an integer. The number $\frac{5}{2}$ is *not* an integer.

Why? The number $\frac{6}{2}$ is 6 divided by 2, which is 3, and 3 is an

integer. The number $\frac{5}{2}$ is 5 divided by 2, which is $2\frac{1}{2}$, and

$2\frac{1}{2}$ is not an integer.

Which Is Greater?

Sometimes mathematicians want to compare two numbers and decide which is larger and which is smaller. But instead of larger and smaller, mathematicians use the words greater than and less than. They say, "Seven is greater than three" or "Three is less than seven." Mathematicians use the symbols ">" and "<" to indicate these relationships.

1+2=3 MATH TALK!

The symbol ">" means "greater than."
For example, 6 > 5 means "six is greater than five."

The symbol "<" means "less than."
For example, 3 < 8 means "three is less than eight."

Here is an example of two inequalities that mean the same thing. Notice that the arrow always points to the smaller number.

$2 > 1$ (Two is greater than one.)

$1 < 2$ (One is less than two.)

Now let's look at some pairs of positive and negative integers, and decide, for each pair, which integer is larger and which is smaller.

- -3 and 5

  Hint: Positive numbers are always greater than negative numbers.

  $-3 < 5$ (Negative three is less than five.)
  Or you could write
  $5 > -3$ (Positive five is greater than negative three.)

- 6 and 0

    Hint: Positive numbers are always greater than zero.
    $6 > 0$ (Six is greater than zero.)
    Or you could write
    $0 < 6$ (Zero is less than six.)

PAINLESS TIP

Mathematical sentences with ">" or "<" are called *inequalities*.

Mathematical sentences with "=" are called *equalities* or *equations*.

- -3 and 0

    Hint: Negative numbers are always less than zero.
    $-3 < 0$ (Negative three is less than zero.)
    Or you could write
    $0 > -3$ (Zero is greater than negative three.)

- -4 and -10

    Hint: The larger a negative number looks, the smaller it actually is.
    $-4 > -10$ (Negative four is greater than negative ten.)
    Or you could write
    $-10 < -4$ (Negative ten is less than negative four.)

CAUTION—Major Mistake Territory!

When comparing two negative numbers, remember that the number that looks larger is actually smaller. Negative six looks larger than negative two, but it is actually smaller. Still confused? Graph both numbers on a number line. The negative number closer to zero is always larger.

BRAIN TICKLERS   Set # 7

Mark each statement true (T) or false (F).

Hint: Five of these statements are true. The sum of the problem numbers of the true statements is 25.

_____ 1. −7 is an integer.

_____ 2. $\frac{3}{2}$ is an integer.

_____ 3. 5 is an integer.

_____ 4. 0 is a positive integer.

_____ 5. 2 < 10   _____ 8. (−3) < (−5)

_____ 6. (−5) > (8)   _____ 9. (−1) > (−4)

_____ 7. (−7) < (−5)   _____ 10. 0 < −1

(Answers are on page 45.)

Adding Integers

When you add integers, the problem may look like one of the four types of problems below.

Case 1: Both numbers are positive.
  Example: $(+3) + (+5)$

Case 2: Both numbers are negative.
  Example: $(-2) + (-4)$

Case 3: One of the two numbers is positive and one is negative.
  Example: $(+3) + (-8)$ or $(-5) + (+7)$

Case 4: One of the two numbers is zero.
  Example: $0 + (+2)$ or $4 + 0$

Now let's see how to solve each of these types of problems.

Case 1: Both numbers are positive.

Painless Solution: Add the numbers just as you would add any two numbers. The answer is always positive.

Examples:

$$(+3) + (+8) = +11$$
$$2 + (+4) = +6$$
$$3 + 2 = 5$$

Case 2: Both numbers are negative.

Painless Solution: Pretend both numbers are positive. Add them. Place a negative sign in front of the answer.

Examples:

$$(-3) + (-8) = -11$$
$$(-2) + (-4) = -6$$
$$-5 + (-5) = -10$$

Case 3: One number is positive and one number is negative.

Painless Solution: To add two numbers together where one is positive and the other negative, find the absolute value of both numbers. Subtract the smaller number from the larger number. Give the answer the sign of the number that has the largest absolute value.

Examples:

$$(-3) + (+8) = ?$$

Find the absolute value of both numbers.

$$\left|-3\right| = 3 \text{ and } \left|+8\right| = 8$$

Subtract the number with the smaller absolute value from the number with the larger absolute value.

Examples:

$$8 - 3 = 5$$

Give the answer the sign of the number with the larger absolute value. Eight has the larger absolute value and 8 was positive, so the answer is positive.

$$(-3) + (+8) = +5$$

$(+2) + (-4) = ?$

Find the absolute value of both numbers.

$$\left|+2\right| = 2 \text{ and } \left|-4\right| = 4$$

Subtract the number with the smaller absolute value from the number with the larger absolute value.

$$4 - 2 = 2$$

Give the answer the sign of the number with the larger absolute value. Negative four has the larger absolute value and negative four is a negative number, so the answer is negative.

$$(+2) + (-4) = -2$$

1+2=3   MATH TALK!

The *absolute value* of any number is positive no matter whether the original number is positive or negative.

The absolute value of 6 is 6, of −4 is 4, and of 0 is 0.

The symbol for the absolute value of a number is a bar on each side of the number. $|-3|$ means the absolute value of negative three. $|9|$ means the absolute value of nine.

Case 4: One of the numbers is zero.

Painless Solution: Zero plus any number is that number.

Examples:

$$(+2) + 0 = +2$$
$$0 + (-8) = -8$$

Visual Clue: Use a number line to help in adding integers. Start at the first number. Then, if the second number is positive, move to the right. If the second number is negative, move to the left.

Example: 3 + (−2) =

Start at 3 and then move to the left two spaces.

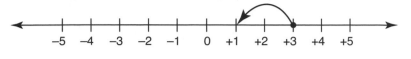

$$3 + (-2) = 1$$

Example: −4 + 2 =

Start at −4 and then move to the right two spaces.

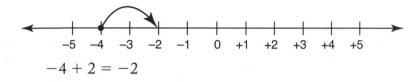

$$-4 + 2 = -2$$

**BRAIN TICKLERS   Set # 8**

Solve the following addition problems.

1. (+5) + 0
2. (+3) + (+6)
3. (−3) + (+6)
4. 0 + (−1)

5. (+4) + (−4)
6. (−6) + (−3)
7. (−5) + (+2)
8. 5 + 2

(Answers are on page 45.)

Subtracting Integers

When you subtract one integer from another integer, there are six possible cases.

Case 1: Both numbers are positive.

　　　　Example: (+5) − (+3)

Case 2: Both numbers are negative.

　　　　Example: (−7) − (−4)

Case 3: The first number is positive and the second is negative.

　　　　Example: (+3) − (−4)

Case 4: The first number is negative and the second number is positive.

Example: $(-5) - (+3)$

Case 5: The first number is zero.

Example: $0 - (-3)$

Case 6: The second number is zero.

Example: $(3) - 0$

Wow! Six possible cases. How do you remember how to subtract one number from another? Just use the *Painless Solution*. Just remember to *keep, change, change*. Keep the first number the same, change the subtraction problem into an addition problem, and change the sign of the last number.

Watch how easy it is to solve these subtraction problems with the *Painless Solution*.

$(+7) - (-3)$

Keep the first number the same and **change** the subtraction problem into an addition problem.
$(+7) + (-3)$
Change the sign of the last number.
$(+7) + (+3)$
Solve the problem.
$(+7) + (+3) = 10$

$(-4) - (-3)$

Keep the first number the same and **change** the subtraction problem into an addition problem.
$(-4) + (-3)$
Change the sign of the last number.
$(-4) + (+3)$
Solve the problem.
$(-4) + (+3) = -1$

$$0 - (+5)$$

Keep the first number the same and **change** the subtraction problem into an addition problem.

$$0 + (+5)$$

Change the sign of the last number.

$$0 + (-5)$$

Solve the problem.

$$0 + (-5) = -5$$

BRAIN TICKLERS   Set # 9

The first step in solving a subtraction problem is to change the subtraction problem into an addition problem. In the lists below, match each subtraction problem on the left to the correct addition problem. If you get all six correct, you will spell a word.

_____ 1. 6 − (−4)   T. 6 + (4)

_____ 2. −6 − (4)   I. 6 + (−4)

_____ 3. 6 − (4)   C. −6 + (4)

_____ 4. (−6) − (−4)   R. −6 + (−4)

_____ 5. 0 − (−4)   Y. 0 + (−4)

_____ 6. 0 − (+4)   K. 0 + 4

(Answers are on page 45.)

BRAIN TICKLERS   Set # 10

Now solve these same subtraction problems. Use the addition problems in Brain Ticklers #9 for help.

1. 6 − (−4)   4. −6 − (−4)

2. −6 − (4)   5. 0 − (−4)

3. 6 − (4)   6. 0 − (+4)

(Answers are on page 45.)

REMINDER

Just change the subtraction problem into an addition problem, and take the opposite of the number being subtracted. Then solve the new addition problem.

Example: 6 − (−4)

6 − (−4) becomes 6 + (−4) when it is changed to an addition problem.

6 + (−4) becomes 6 + (+4) after the sign of the last number is changed.

6 + (+4) = 10

Multiplying Integers

Multiplying integers is easy. When you multiply two integers, there are four possible cases.

Case 1: Both numbers are positive.

Example: $6 \cdot 4$

Case 2: Both numbers are negative.

Example: $(-3)(-2)$

Case 3: One number is positive and the other is negative.

Example: $(-5)(+2)$ or $(+8)(-3)$

Case 4: One of the two numbers is zero.

Example: $(-6) \cdot 0$ or $(0)(5)$

Here is how you solve multiplication problems with integers.

Case 1: Both numbers are positive.

Painless Solution: Just multiply the numbers. The answer is always positive.

Examples:

$$5 \times 3 = 15$$
$$(+6)(+4) = (+24)$$
$$2(9) = (18)$$

Case 2: Both numbers are negative.

Painless Solution: Just pretend the numbers are positive. Multiply the numbers. The answer is always positive.

Examples:

$$(-5)(-3) = 15$$
$$-6(-4) = 24$$

Case 3: One number is positive and the other negative.

Painless Solution: Just pretend the numbers are positive. Multiply the numbers together. The answer is always negative.

Examples:

$$(-4)(+3) = (-12)$$
$$5 \times (-2) = (-10)$$

Case 4: One of the two numbers is zero.

Painless Solution: The answer is always zero. It doesn't matter whether you are multiplying a positive number by zero or a negative number by zero. Zero times any number or any number times zero is always zero.

Examples:

$$0 \times 7 = 0$$
$$(-8) \times 0 = 0$$
$$(+4) \times 0 = 0$$
$$0(-1) = 0$$

PAINLESS TIP

A positive number times a positive number is a positive number.

A negative number times a negative number is a positive number.

A positive number times a negative number is a negative number.

If there is no sign in front of a number, it is positive.

BRAIN TICKLERS   Set # 11

Solve the following multiplication problems.

1. (−2)(−8)   4. 5 · 0

2. (3)(−3)   5. 3 · 3

3. (8)(−2)

(Answers are on page 45.)

Dividing Integers

When you divide two integers, there are five possible cases.

Case 1: Both numbers are positive.

Example: 21 ÷ 7

Case 2: Both numbers are negative.

Example: (−15) ÷ (−3)

Case 3: One number is negative and one number is positive.

Example: (+8) ÷ (−4)

Case 4: The dividend is zero.

Example: 0 ÷ (−2)

Case 5: The divisor is zero.

> Example: $6 \div 0$

Here is how you solve each of these cases.

Case 1: Both numbers are positive.

Painless Solution: Divide the numbers. The answer is always positive.

Examples:

$$8 \div 2 = 4$$
$$(+12) \div (+4) = +3$$

Case 2: Both numbers are negative.

Painless Solution: Pretend both numbers are positive. Divide the numbers. The answer is always positive.

Examples:

$$-8 \div (-2) = +4$$
$$-15 \div (-3) = +5$$

Case 3: One number is positive and the other negative.

Painless Solution: Pretend the numbers are positive. Divide the numbers. The answer is always negative.

Examples:

$$-9 \div 3 = -3$$
$$15 \div (-3) = -5$$

Case 4: The dividend is zero.

Painless Solution: Zero divided by any number (except 0) is zero. The answer is always zero.

Examples:

$$0 \div 6 = 0$$
$$0 \div (-3) = 0$$

Case 5: The divisor is zero.

Painless Solution: Division by zero is always undefined.

Examples:

$$4 \div 0 = \text{undefined}$$
$$-8 \div 0 = \text{undefined}$$

PAINLESS TIP

The rules for division are the same as those for multiplication.

A positive number divided by a positive number is positive.

A negative number divided by a negative number is positive.

A positive number divided by a negative number is negative.

A negative number divided by a positive number is negative.

BRAIN TICKLERS   Set # 12

Solve these division problems.

1. $5 \div 5$

2. $5 \div (-5)$

3. $(-6) \div 3$

4. $0 \div 10$

5. $(-6) \div (-3)$

6. $10 \div 0$

(Answers are on page 46.)

SUPER BRAIN TICKLERS

See if you can figure out the sign of the correct answer to each problem. Then circle the letter next to the correct answer and spell a phrase.

1. $(-6) + (-2) =$ $(-)$ E or $(+)$ S

2. $(-12) \div (-3) =$ $(-)$ O or $(+)$ A

3. $(-6) - (-2) =$ $(-)$ S or $(+)$ T

4. $-7 \div (+1) =$ $(-)$ Y or $(+)$ N

5. $2 - (-2) =$ $(-)$ I or $(+)$ A

6. $(-3)(-2) =$ $(-)$ T or $(+)$ S

7. $(-4) + (-7) =$ $(-)$ P or $(+)$ L

8. $(-6) - (-8) =$ $(-)$ R or $(+)$ I

9. $4 \times (-2) =$ $(-)$ E or $(+)$ R

(Answers are on page 46.)

Word Problems

Many people go "Ugh!" when they hear that it's time to do word problems. Word problems can be painless. To solve a word problem, all you have to do is change Plain English into Math Talk. Here is how to solve a few word problems that use integers.

Problem 1: An elevator went up three floors and down two floors. How much higher or lower was the elevator than when it started?

Painless Solution:
The elevator went up three floors $(+3)$.
The elevator went down two floors (-2).
The word *and* means "add" $(+)$.
The problem is $(+3) + (-2)$.
The answer is $(+3) + (-2) = +1$.
The elevator was one floor higher than when it started.

Problem 2: Today's high temperature was six degrees. Today's low temperature was two degrees below zero. What was the change in temperature?

> *Painless Solution:*
> The high temperature was six degrees ($+6$).
> The low temperature was two degrees below zero (-2).
> The word *change* means "subtract" ($-$).
> The problem is ($+6$) $-$ (-2).
> The answer is ($+6$) $-$ (-2) $= +8$.
> There was an eight-degree change in temperature.

Problem 3: The temperature dropped two degrees every hour. How many degrees did it drop in six hours?

> *Painless Solution:*
> The temperature dropped two degrees (-2).
> The time that elapsed was six hours ($+6$).
> Type of problem: multiplication.
> The problem is (-2)($+6$).
> The answer is (-2)($+6$) $= -12$.
> The temperature dropped 12 degrees.

Problem 4: Bob spends $3 a day on lunch. So far this week Bob has spent a total of $12 for lunches. For how many days has Bob bought lunch?

> *Painless Solution:*
> The total spent for lunches was $12.
> The cost of lunch for one day was $3.
> Type of problem: division.
> The problem is 12 $\div$ 3.
> The answer is $12 $\div$ $3 = 4$.
> Bob bought lunch for four days.

BRAIN TICKLERS—THE ANSWERS
Set # 7, page 32

1. T   3. T   5. T   7. T   9. T
2. F   4. F   6. F   8. F   10. F

Notice that $1 + 3 + 5 + 7 + 9 = 25$.

Set # 8, page 35

1. $(+5) + 0 = +5$
2. $(+3) + (+6) = +9$
3. $(-3) + (+6) = +3$
4. $0 + (-1) = -1$
5. $(+4) + (-4) = 0$
6. $(-6) + (-3) = -9$
7. $(-5) + (+2) = -3$
8. $5 + 2 = +7$

Set # 9, page 37

TRICKY

Set # 10, page 37

1. $6 - (-4) = 10$
2. $-6 - (4) = -10$
3. $6 - (4) = 2$
4. $-6 - (-4) = -2$
5. $0 - (-4) = 4$
6. $0 - (+4) = -4$

Set # 11, page 40

1. $(-2)(-8) = 16$
2. $(3)(-3) = -9$
3. $(8)(-2) = -16$
4. $5 \cdot 0 = 0$
5. $3 \cdot 3 = 9$

Set # 12, page 42

1. $5 \div 5 = 1$
2. $5 \div (-5) = -1$
3. $(-6) \div 3 = -2$
4. $0 \div 10 = 0$
5. $(-6) \div (-3) = 2$
6. $10 \div 0$ is undefined.

Super Brain Ticklers, page 43

EASY AS PIE

Chapter 3

Solving Equations with One Variable

Defining the Terms

An *equation* is a mathematical sentence with an equals sign in it. A *variable* is a letter that is used to represent a number. Some equations have variables in them, and some do not.

$3 + 5 = 8$ is an equation.

$3x + 1 = 4$ is an equation.

$2x + 7x + 1$ is *not* an equation. It does not have an equals sign.

$3 + 2 > 5$ is a mathematical sentence, but it is not an equation because it does not have an equals sign.

All equations are mathematical sentences. But not all mathematical sentences are equations.

1+2=3 MATH TALK!

Think of the letter x as a mystery number. Here is how to change the following equations from Math Talk into Plain English.

$$x + 3 = 7$$
A mystery number plus three equals seven.

$$2x - 4 = 8$$
Two times a mystery number minus four equals eight.

$$\frac{1}{2}x = 5$$

One half of a mystery number equals five.

$$3(x + 1) = 0$$
Three times the quantity of a mystery number plus one equals zero.

One of the major goals of algebra is to figure out the value of the mystery number. When you figure out the value of the mystery number and insert it into the equation, the mathematical sentence will be true.

Sometimes you can look at an equation and guess the value of the mystery number.

Look at the equation $x + 1 = 2$. What do you think the mystery number is? You're right, it's one: $1 + 1 = 2$.

What do you think a stands for in the equation $2a = 10$? You're right; it's five: $2(5) = 10$.

Sometimes you can look at an equation and figure out the correct answer, but most of the time you have to solve an equation using the principles of algebra. Could you solve the equation $3(x + 2) + 5 = 6(x - 1) - 4$ in your head? Probably not.

But by the time you finish this chapter, you will consider the equation as easy as pie.

Solving Equations

Solving equations is *painless*. There are three steps to solving an equation with one variable.

Step 1: Simplify each side of the equation.

Step 2: Add and/or subtract.

Step 3: Multiply or divide.

Step 1: Simplify each side of the equation

To simplify an equation, first simplify the left side. Next, simplify the right side. When you simplify each side of the equation, use the Order of Operations.

> **Simplify: $5x = 3(4 + 1)$**
> $5x = 3(4 + 1)$
> $5x = 3(5)$
> $5x = 15$

Simplify: $5(x + 2) = 10 + 5$

First simplify the left side of the equation.

$$5(x) + 5(2) = 10 + 5$$
$$5x + 10 = 10 + 5$$

Next simplify the right side of the equation.

$$5x + 10 = 15$$

Simplify: $4x + 2x - 7 + 9 + x = 5x - x$

First, simplify the left side of the equation.

$$7x + 2 = 5x - x$$

Next simplify the right side of the equation. Subtract x from $5x$.

$$7x + 2 = 4x$$

CAUTION—Major Mistake Territory!

In simplifying an equation so that it can be solved:

Whatever is on the left side of the equals sign stays on the left side of the equals sign.

Whatever is on the right side of the equals sign stays on the right side of the equals sign.

Don't mix the terms on the left side of the equation with the terms on the right side of the equation.

BRAIN TICKLERS   Set # 13

Simplify these equations. Remember to Please Excuse My Dear Aunt Sally.

1. $3(x + 2) = 0$
2. $5x + 1 + 2x = 4$
3. $2(x + 1) + 2x = 8$
4. $5x - 2x = 3(4 + 1)$
5. $x - 4x = 5 - 2 - 8$
6. $3x - 2x + x - 4 + 3 - 2 = 0$

(Answers are on page 60.)

Step 2: Use addition and subtraction to solve equations with one variable

Once an equation is simplified, the next step is to get all the variables on one side of the equation and all the numbers on the other side. To do this, you add and/or subtract the same number or variable from both sides of the equation.

$x - 4 = 8$

Find the value of x.

The only variable is already on the left side of the equation. To move the four to the right side of the equation, add four to both sides of the equation. When you add 4 to the left side of the equation, the -4 will disappear. The only variable will be on the left side of the equation, and the two numbers will be on the right side of the equation.

$x - 4 + 4 = 8 + 4$

Combine terms.

$x = 12$

$x + 5 = 12$

Find the value of x.

The only variable is already on the left side of the equation. To get all the numbers on the right side of the equation, subtract five from both sides of the equation.

$x + 5 - 5 = 12 - 5$

Combine terms.

$x = 7$

$4x - 5 = 3x - 1$

Find the value of x.

Subtract $3x$ from both sides of the equation so that only one x remains on the left of the equals sign.

$4x - 5 - 3x = 3x - 1 - 3x$

Simplify.

$x - 5 = -1$

Add five to both sides of the equation.

$x - 5 + 5 = -1 + 5$

Simplify to find the solution.

$x = 4$

REMINDER

Remember: Whatever you do to one side of an equation, you must do to the other side of the equation. You must treat both sides equally.

BRAIN TICKLERS   Set # 14

Solve each equation by adding or subtracting the same number or variable from both sides of the equation. Keep the variable x on the left side of the equation and the numbers on the right side of the equation.

1. $x - 3 = 10$
2. $x - (-6) = 12$
3. $-5 + x = 4$
4. $x + 7 = 3$
5. $5 + 2x = 2 + x$
6. $4x - 4 = -6 + 3x$

(Answers are on page 60.)

Step 3: Use multiplication/division to solve equations

You can multiply one side of an equation by any number as long as you multiply the other side of the equation by the same number. Both sides of the equation will still be equal. You can also divide one side of an equation by any number as long as you divide the other side of the equation by the same number.

Once you have all the variables on one side of the equation and all the numbers on the other side of the equation, how do you decide what number to multiply or divide by? You pick the number that will give you only one x.

If the equation has $5x$, divide both sides by 5.

If the equation has $\frac{1}{2}x$, multiply both sides by $\frac{2}{1}$.

If the equation has $\frac{3}{5}x$, multiply both sides by $\frac{5}{3}$.

$4x = 8$

Find the value of x.
Divide both sides of the equation by 4.

$$4x \div 4 = 8 \div 4$$

This is the solution.

$$x = 2$$

$-2x = -10$

Find the value of x.
Divide both sides of the equation by -2 so that a positive x remains on the left side of the equation.

$$-2x \div -2 = -10 \div -2$$

This is the solution.

$$x = 5$$

$\frac{1}{2}x = 10$

Find the value of x.
Multiply both sides of the equation by 2.

$$2\left(\frac{1}{2}x\right) = 2(10)$$

This is the solution.

$$x = 20$$

$-\frac{5}{2}x = 10$

Find the value of x.
Multiply both sides of the equation by the inverse of $-\frac{5}{2}$.

$$\left(-\frac{2}{5}\right)\left(-\frac{5}{2}x\right) = \left(-\frac{2}{5}\right)(10)$$

This is the solution.

$$x = -4$$

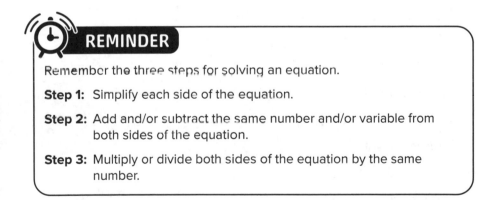

BRAIN TICKLERS   Set # 15

Solve these equations.

1. $\frac{1}{3}x = 2$ 4. $-\frac{1}{4}x = \frac{1}{2}$

2. $3x = -3$ 5. $\frac{3}{2}x = -3$

3. $4x = 0$

(Answers are on page 61.)

Solving More Equations

REMINDER

Remember the three steps for solving an equation.

Step 1: Simplify each side of the equation.

Step 2: Add and/or subtract the same number and/or variable from both sides of the equation.

Step 3: Multiply or divide both sides of the equation by the same number.

Here is an example of an equation to solve where you must use all three steps.

$$3x - x + 3 = 7$$

Simplify the equation by combining like terms.

$$(3x - x) + 3 = 7$$
$$2x + 3 = 7$$

Subtract 3 from both sides of the equation.

$$2x + 3 - 3 = 7 - 3$$
$$2x = 4$$

Divide both sides of the equation by 2.

$$2x \div 2 = 4 \div 2$$

This is the solution.

$$x = 2$$

Here is another example. Remember to use the three steps.

$3(x + 5) = 15 + 6$

Distribute the 3 in front of the expression $(x + 5)$.

$$3(x + 5) = 15 + 6$$
$$3x + 15 = 15 + 6$$

Subtract 15 from both sides of the equation.

$$3x + 15 - 15 = 15 + 6 - 15$$
$$3x = 6$$

Divide both sides of the equation by 3.

$$3x \div 3 = 6 \div 3$$

This is the solution.

$$x = 2$$

Here is a third example.

$x - \dfrac{1}{2}x + 5 - 3 = 0$

Simplify the equation by combining like terms.

$$x - \frac{1}{2}x + 5 - 3 = 0$$

$$\frac{1}{2}x + 2 = 0$$

Multiply both sides of the equation by 2. Remember to distribute the 2.

$$2\left(\frac{1}{2}x\right) + 2(2) = 0$$

$$x + 4 = 0$$

Subtract 4 from both sides of the equation.

$$x + 4 - 4 = 0 - 4$$

This is the solution.

$$x = -4$$

BRAIN TICKLERS Set # 16

Solve each of the following equations.

1. $3(x + 1) = 6$

2. $3x - 5x + x = 3 - 2x$

3. $5x + 3 + x = 3 - 6$

4. $\frac{1}{2}x + 5 = 6 - 2$

5. $\frac{2}{3}x + 1 = -5$

6. $5(2x - 2) = 3(x - 1) + 7$

(Answers are on page 61.)

Checking your work

Once you solve an equation, it is important to check your work. Substitute the answer in the original equation wherever you see an x or other variable. Then compute the value of the sentence with the number in place of the variable. If the two sides of the equation are equal, the answer is correct.

Example: Bill solved the equation $3x + 1 = 10$. He came up with $x = 3$.

To check, substitute 3 for x.
$$3(3) + 1 = 10$$

Compute.
$$9 + 1 = 10$$
$$10 = 10$$

Bill was right; $x = 3$.

Example: Jodie solved the equation $3x - 2x + 5 = 2$. She came up with $x = -3$.

To check, substitute -3 for x.
$$3(-3) - 2(-3) + 5 = 2$$

Compute.
$$-9 - (-6) + 5 = 2$$
$$-9 + 6 + 5 = 2$$
$$2 = 2$$

Jodie was right; $x = -3$.

PAINLESS TIP

When you check a problem, you find out whether your answer is right or wrong. To find the correct answer, you must solve the problem again.

Example: Mike solved the equation $3(x + 2) = \frac{1}{2}(x - 2)$. He came up with $x = 4$.

To check, substitute 4 for x.

$$3(4 + 2) = \frac{1}{2}(4 - 2)$$

Compute.

$$3(6) = \frac{1}{2}(2)$$
$$18 = 1$$

Mike was not correct; x is not equal to 4. 18 is not equal to 1.

BRAIN TICKLERS   Set # 17

Which three of the following were solved incorrectly? To find out, check each problem by substituting the answer for the variable.

1. $x + 7 = 10$ $x = 4$

2. $4x = 20$ $x = 5$

3. $2(x - 6) = 0$ $x = 0$

4. $3x + 5 = -4$ $x = -3$

5. $\frac{2}{3}x + 1 = -5$ $x = -9$

6. $4x - 2x - 7 = -1$ $x = -1$

(Answers are on page 62.)

Word Problems

Solving word problems is simply a matter of knowing how to change Plain English into Math Talk! Once you translate a problem correctly, solving word problems is easy.

1+2=3 MATH TALK!

These simple rules should help you change Plain English into Math Talk.

Rule 1: Change the word *equals* or any of the words *is*, *are*, *was*, and *were* into an equals sign.

Rule 2: Use the letter x to represent the phrase "a number."

Let x = a number. Or use the letter x for what you don't know.

BRAIN TICKLERS   Set # 18

Change the following Plain English phrases into Math Talk.

1. five less than a number

2. three more than a number

3. four times a number

4. one fifth of a number

5. the difference between a number and three

6. the product of eight and a number

7. the sum of four and a number

(Answers are on page 63.)

Here is how to change word problems into equations.

Problem 1: A number plus three is twelve. Find the number.

Change this sentence into Math Talk.
Change "a number" to "x."
Change "plus three" to "$+3$."
Change "is" to "$=$."
Change "twelve" to "12."
$x + 3 = 12$

Problem 2: Four times a number plus two is eighteen. Find the number.

Change this sentence into Math Talk.
Change "a number" to "x."
Change "four times a number" to "$4x$."
Change "plus two" to "$+2$."
Change "is" to "$=$."
Change "eighteen" to "18."
$4x + 2 = 18$

Problem 3: Two times the larger of two consecutive integers is three more than three times the smaller integer. Find both integers.

Change this sentence into Math Talk.

Clue: Two consecutive integers are x and $x + 1$; x is the smaller integer and $x + 1$ is the larger integer.

Change "two times the larger of two consecutive integers" to "$2(x + 1)$."
Change "is" to "$=$."
Change "three more" to "$3 +$."
Change "three times the smaller integer" to "$3x$."
$2(x + 1) = 3 + 3x$

SUPER BRAIN TICKLERS

Solve for x.

1. $4x - (2x - 3) = 0$

2. $5(x - 2) = 6(2x + 1)$

3. $4x - 2x + 1 = 5 + x - 7$

4. $\frac{1}{2}x = \frac{1}{4}x + 2$

5. $6(x - 2) - 3(x + 1) = 4(3 + 2)$

(Answers are on page 63.)

BRAIN TICKLERS—THE ANSWERS
Set # 13, page 49

1. $3x + 6 = 0$ 3. $4x + 2 = 8$ 5. $-3x = -5$

2. $7x + 1 = 4$ 4. $3x = 15$ 6. $2x - 3 = 0$

Set # 14, page 51

1. $$x - 3 = 10$$
$$x - 3 + 3 = 10 + 3$$
$$x = 13$$

2. $$x - (-6) = 12$$
$$x - (-6) + (-6) = 12 - 6$$
$$x = 6$$

3. $$-5 + x = 4$$
$$-5 + 5 + x = 4 + 5$$
$$x = 9$$

4. $$x + 7 = 3$$
$$x + 7 - 7 = 3 - 7$$
$$x = -4$$

5. $$5 + 2x = 2 + x$$
$$5 - 5 + 2x - x = 2 - 5 + x - x$$
$$x = -3$$

6. $$4x - 4 = -6 + 3x$$
$$4x - 3x - 4 + 4 = -6 + 4 + 3x - 3x$$
$$x = -2$$

Set # 15, page 53

1. $\frac{1}{3}x = 2$

   $3\left(\frac{1}{3}x\right) = 3(2)$

   $x = 6$

4. $-\frac{1}{4}x = \frac{1}{2}$

   $-4\left(-\frac{1}{4}x\right) = -4\left(\frac{1}{2}\right)$

   $x = -2$

2. $3x = -3$

   $\frac{3x}{3} = \frac{-3}{3}$

   $x = -1$

5. $\frac{3}{2}x = -3$

   $\frac{2}{3}\left(\frac{3}{2}x\right) = \frac{2}{3}(-3)$

   $x = -2$

3. $4x = 0$

   $\frac{4x}{4} = \frac{0}{4}$

   $x = 0$

Set # 16, page 55

1. $3(x + 1) = 6$

   $3x + 3 = 6$

   $3x = 3$

   $x = 1$

2. $3x - 5x + x = 3 - 2x$

   $-x = 3 - 2x$

   $x = 3$

3. $5x + 3 + x = 3 - 6$

   $6x + 3 = -3$

   $6x = -6$

   $x = -1$

4. $\frac{1}{2}x + 5 = 6 - 2$

   $\frac{1}{2}x + 5 = 4$

   $\frac{1}{2}x = -1$

   $x = -2$

5. $\frac{2}{3}x + 1 = -5$

$$\frac{2}{3}x = -6$$

$$x = -9$$

6. $5(2x - 2) = 3(x - 1) + 7$
   $10x - 10 = 3x - 3 + 7$
   $10x - 10 = 3x + 4$
   $7x = 14$
   $x = 2$

Set # 17, page 57

1. $x + 7 = 10; x = 4$
   $4 + 7 = 10$
   $11 = 10$

   This problem is solved incorrectly.

2. $4x = 20; x = 5$
   $4(5) = 20$
   $20 = 20$

Correct.

3. $2(x - 6) = 0; x = 0$
   $2(0 - 6) = 0$
   $2(-6) = 0$
   $-12 = 0$

This problem is solved incorrectly.

4. $3x + 5 = -4; x = -3$
   $3(-3) + 5 = -4$
   $-9 + 5 = -4$
   $-4 = -4$

Correct.

5. $\frac{2}{3}x + 1 = -5; x = -9$

$$\frac{2}{3}(-9) + 1 = -5$$

$$-6 + 1 = -5$$

Correct.

6. $4x - 2x - 7 = -1; x = -1$

$$4(-1) - 2(-1) - 7 = -1$$

$$-4 + 2 - 7 = -1$$

$$-9 = -1$$

This problem is solved incorrectly.

Set # 18, page 58

1. $x - 5$

2. $x + 3$

3. $4x$

4. $\frac{1}{5}x$ or $\frac{x}{5}$

5. $x - 3$

6. $8x$

7. $4 + x$ or $x + 4$

Super Brain Ticklers, page 59

1. $x = -\frac{3}{2}$

2. $x = -\frac{16}{7}$

3. $x = -3$

4. $x = 8$

5. $x = \frac{35}{3}$

Chapter 4

Solving Inequalities

An inequality is a sentence in which one side of the expression is greater than or less than the other side of the expression. Inequalities are represented by four different symbols.

> means "greater than."

< means "less than."

≥ means "greater than or equal to."

≤ means "less than or equal to."

Notice that the symbol for greater than or equal to, ≥, is just the symbol for greater than with half an equals sign on the bottom. Similarly, the symbol for less than or equal to, ≤, is just the symbol for less than with half an equals sign on the bottom. Notice also that the symbol ≥ is read as "greater than *or* equal to," not "greater than *and* equal to." No number can be greater than and equal to another number at the same time.

1+2=3 MATH TALK!

Here is how you change the following inequalities from Math Talk into Plain English.

$$4 > 0$$

Four is greater than zero.

$$-3 \geq -7$$

Negative three is greater than or equal to negative seven.

$$2 < 5$$

Two is less than five.

$$-6 \leq 4$$

Negative six is less than or equal to four.

$$8 \leq 8$$

Eight is less than or equal to eight.

An inequality can be true or false.

The inequality $3 > 1$ is true, since three is greater than one.

The inequality $5 < 10$ is true, since five is less than ten.

The inequality $-6 \leq -6$ is true because negative six is equal to negative six.

The inequality $5 > 10$ is false, since five is not greater than ten.

The inequality $-6 \leq -9$ is false, since negative six is not less than negative nine and negative six is not equal to negative nine.

Sometimes one side of an inequality has a variable. The inequality states whether the variable is larger, smaller, or maybe even equal to a specific number.

BRAIN TICKLERS   Set # 19

Is each of the following statements true or false?

_____ 1. $3 > 4$ _____ 4. $-6 \leq -1$

_____ 2. $6 \geq 6$ _____ 5. $0 \leq 0$

_____ 3. $0 < -4$

(Answers are on page 79.)

1+2=3 MATH TALK!

Here is how you change the following inequalities from Math Talk into Plain English.

$$x > 4$$

A mystery number is greater than four.

$$x < -1$$

A mystery number is less than negative one.

$$x \leq -2$$

A mystery number is less than or equal to negative two.

$$x \geq 0$$

A mystery number is greater than or equal to zero.

If an inequality has a variable in it, some numbers will make this inequality true while others will make it false. For example, consider $x > 2$. If x is equal to 3, 4, or 5, this inequality is true.

It is even true if x is equal to $2\frac{1}{2}$. But $x > 2$ is false if x is equal to 0, -1, or -2. The inequality $x > 2$ is false even if x is equal to two, since two is not greater than two.

The $x > 1$ example (x is greater than one) means that x can be any number greater than one, not just any whole number greater than one. If $x > 1$, then x can be 1.1 or 1.234 or 1,000,001.5.

Graphing Inequalities

Often, inequalities with variables are graphed. The graph gives you a quick picture of all the mystery numbers on the number line that will work. There are two steps to graphing an inequality on the number line.

Step 1: Locate the number in the inequality on the number line. If the inequality is either $>$ or $<$, circle the number. If the inequality is either $\geq$ or $\leq$, circle and shade the number.

Step 2: Place x on the left side of the inequality. If the inequality is either $>$ or $\geq$, shade the number line to the right of the number. If the inequality is either $<$ or $\leq$, shade the number line to the left of the number.

Now let's try an example.

$x > 2$

Step 1: Locate the number in the inequality on the number line.

Two is marked on the number line. If the inequality is either $>$ or $<$, circle the number.

If the inequality is either $\geq$ or $\leq$, circle and shade the number.

Circle the number two. Circling the number means that it is not included in the graph.

Step 2: Place the variable on the left side of the inequality. If the inequality is either > or ≥, shade the number line to the right of the number.

Since the inequality is $x > 2$, shade the number line to the right of the number two. Notice that all the numbers are shaded, not just all the whole numbers. All the numbers to the right of two are greater than two.

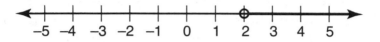

Here is another example:

$x \leq -1$

Step 1: Locate the number in the inequality on the number line. Because the inequality is $x \leq -1$, circle and shade -1. Circling and shading -1 means that -1 is included in the graph.

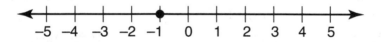

Step 2: Because the graph is $x \leq -1$, shade the number line to the left of -1.

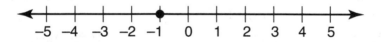

1+2=3 **MATH TALK!**

Here is how you read the following graphs and change them from Math Talk into Plain English.

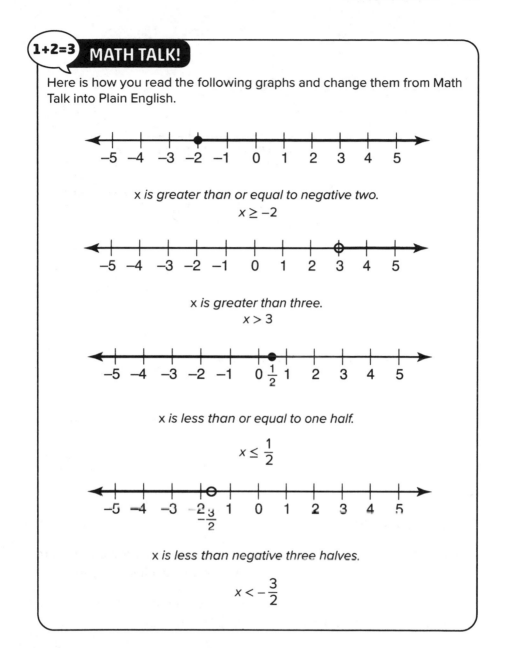

x is greater than or equal to negative two.

$$x \geq -2$$

x is greater than three.

$$x > 3$$

x is less than or equal to one half.

$$x \leq \frac{1}{2}$$

x is less than negative three halves.

$$x < -\frac{3}{2}$$

BRAIN TICKLERS   Set # 20

Graph the following inequalities.

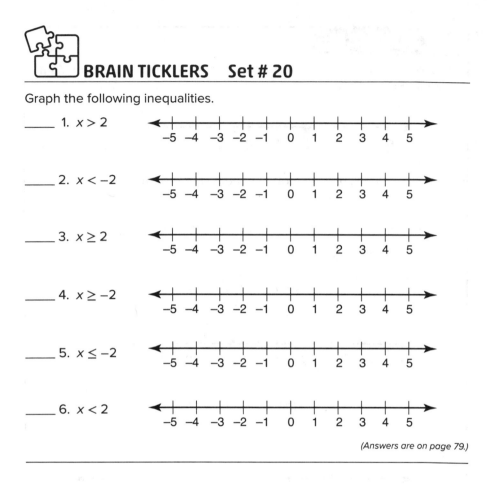

_____ 1. $x > 2$

_____ 2. $x < -2$

_____ 3. $x \geq 2$

_____ 4. $x \geq -2$

_____ 5. $x \leq -2$

_____ 6. $x < 2$

(Answers are on page 79.)

Solving Inequalities

Solving inequalities is *almost* exactly like solving equations. There are a couple of important differences, however, so pay close attention.

Follow the same three steps when solving inequalities that you followed when solving equations, and add a fourth step.

Step 1:   Simplify each side of the inequality. Simplifying each side of the inequality is a two-step process. First simplify the left side. Next simplify the right side.

Step 2:   Add and/or subtract numbers and/or variable terms from both sides of the inequality. Move all the variables to one side of the inequality and all the numbers to the other side.

Step 3: Multiply or divide both sides of the inequality by the same number. *If you multiply or divide by a negative number, reverse the direction of the inequality.*

Step 4: Graph the answer on the number line.

Now you can solve an inequality. Here are three examples.

$2(x - 1) > 4$

Step 1: Simplify the left side of the inequality.

The inequality is now $2x - 2 > 4$.

Step 2: Add or subtract the same number to or from both sides of the inequality.

Add 2 to both sides of the inequality.

$2x - 2 + 2 > 4 + 2$

Simplify.

$2x > 6$

Step 3: Multiply or divide both sides of the inequality by the same number.

Divide both sides of the inequality by 2.

$$\frac{2x}{2} > \frac{6}{2}$$

$x > 3$

Step 4: Graph the answer on the number line.

Circle the number 3 on the number line. Circling 3 indicates that 3 is not included in the graph.

Shade the number line to the right of the number 3.

All the numbers greater than 3 will make the inequality $2(x - 1) > 4$ true.

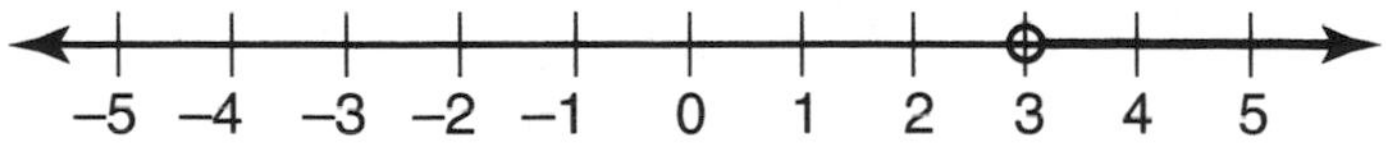

$2x - 5x + 4 \leq 10$

Step 1: Simplify each side of the inequality.

$-3x + 4 \leq 10$

Step 2: Add or subtract the same number to or from both sides of the inequality.

Subtract 4 from both sides.
$-3x + 4 - 4 \leq 10 - 4$
Simplify.
$-3x \leq 6$

Step 3: Multiply or divide both sides of the inequality by the same number.

Divide both sides by -3. Why -3? Because you want to have only one positive x on the left side of the inequality. *Because you are dividing by a negative number, you must reverse the direction of the inequality.*

$\dfrac{-3x}{-3} \geq \dfrac{6}{-3}$

Simplify.

$x \geq -2$

Step 4: Graph the answer on the number line.

Circle the number -2 on the number line. Shade the circle. Circling and shading -2 indicates that it is included in the graph.

Shade the number line to the right of the number -2, since the inequality states that x is greater than -2.

All the numbers greater than or equal to -2 will make the inequality $2x - 5x + 4 \leq 10$ true.

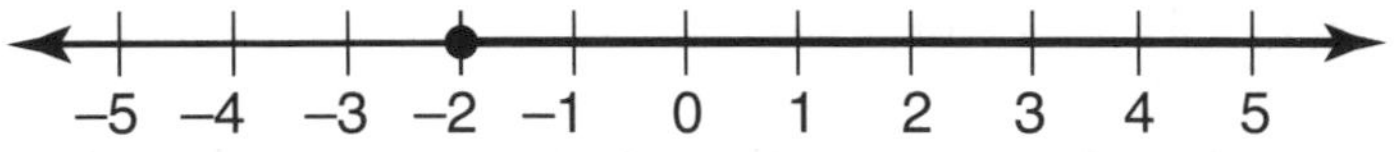

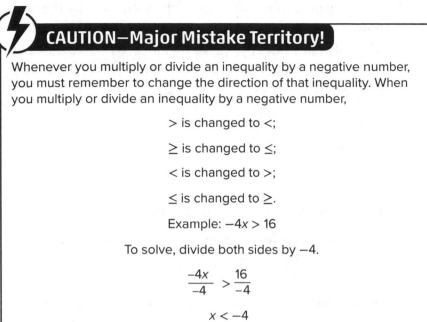

$$-\frac{1}{2}(x - 10) \geq 7$$

Step 1: Simplify each side of the inequality.

Simplify the left side. Multiply $(x - 10)$ by $-\frac{1}{2}$.

Since you are multiplying only one side of the inequality by a negative number $\left(-\frac{1}{2}\right)$, *don't* reverse the inequality sign.

$$-\frac{1}{2}x + 5 \geq 7$$

Step 2: Add or subtract the same number to or from both sides of the inequality.

Subtract 5 from both sides of the inequality.

$$-\frac{1}{2}x + 5 - 5 \geq 7 - 5$$

Compute.

$$-\frac{1}{2}x \geq 2$$

Step 3: Multiply or divide both sides of the inequality by the same number.

Multiply both sides by -2. Remember to reverse the direction of the inequality, since you are multiplying both sides by a negative number.

$$-2\left(-\frac{1}{2}x\right) \le -2(2)$$

Compute.

$$x \le -4$$

Step 4: Graph the answer on the number line.

All the numbers less than or equal to -4 make the inequality $-\frac{1}{2}(x-10) \ge 7$ true.

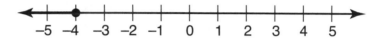

PAINLESS TIP

When solving an inequality with one variable, follow these steps to success.

Step 1: Simplify both sides of the inequality.

Step 2: Add or subtract the same number or variable on both sides of the inequality. Make sure all of the variables are on one side of the inequality and all the numbers are on the other side of the inequality.

Step 3: Multiply or divide both sides of the inequality by the same number. If you multiply or divide by a negative number, you must reverse the direction of the inequality.

BRAIN TICKLERS   Set # 21

Solve these inequalities.

1. $3x + 5 > 7$

2. $-\frac{1}{2}x - 2 > 8$

3. $4(x - 2) < 8$

4. $-5(x - 1) < 5(x + 1)$

5. $4x + 2 - 4x \geq 5 - x - 4$

6. $\frac{1}{3}x - 2 \leq \frac{2}{3}x - 6$

(Answers are on page 79.)

Checking your work

To check whether you solved an inequality correctly, follow these simple steps.

Step 1: Change the inequality sign in the problem to an equals sign.

Step 2: Substitute the number in the answer for the variable.

If the sentence is true, continue to Step 3.

If the sentence is not true, STOP. The answer is wrong.

Step 3: Substitute zero in the inequality to check the direction of the inequality.

Example: The problem was $x + 3 > 4$. Allison thinks the answer is $x > 1$. Check to see whether Allison is right or wrong.

Step 1: Change the inequality into an equation. $x + 3 > 4$ becomes $x + 3 = 4$.

Step 2: Substitute the answer for x.

Substitute 1 for x.
$1 + 3 = 4$
This is a true sentence.

Step 3: Check the direction of the inequality by substituting 0 for x.
$0 + 3 > 4$
This is not true, so zero is not part of the solution set. If it were, the answer would be $x < 1$. But because zero is not part of the solution, the correct answer is $x > 1$. Allison was correct.

Word Problems

The trickiest part of solving word problems with inequalities is changing the problems from Plain English to Math Talk. Much of an inequality problem is changed the same way you change a word problem into an equation. The trick is deciding which inequality to use and which way it should point. Here are some tips that should help.

Look for these phrases when solving inequalities.

Whenever the following phrases are used, insert $>$.

. . . is more than . . .
. . . is bigger than . . .
. . . is greater than . . .
. . . is larger than . . .

Whenever the following phrases are used, insert $<$.

. . . is less than . . .
. . . is smaller than . . .

Whenever the following phrases are used, insert $\geq$.

. . . is greater than or equal to . . .
. . . is at least . . .

Whenever the following phrases are used, insert $\leq$.

... is less than or equal to ...
... is at most ...

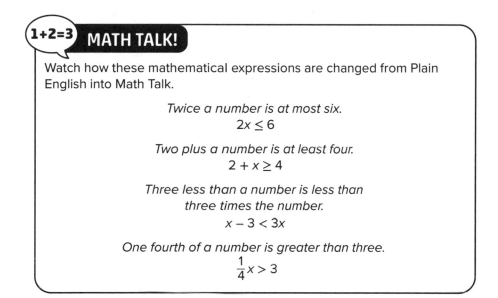

1+2=3 MATH TALK!

Watch how these mathematical expressions are changed from Plain English into Math Talk.

Twice a number is at most six.
$2x \leq 6$

Two plus a number is at least four.
$2 + x \geq 4$

Three less than a number is less than three times the number.
$x - 3 < 3x$

One fourth of a number is greater than three.
$\frac{1}{4}x > 3$

Here are two word problems that are solved. Study each of them.

Problem 1: Three times a number plus one is at most ten. What could the number be?

First change the problem from Plain English into Math Talk.
Change "three times a number" to "$3x$."
Change "plus one" to "$+1$."
Change "is at most" to "$\leq$."
Change "ten" to "10."
In Math Talk, the problem now reads $3x + 1 \leq 10$.

Now solve the inequality.
Subtract 1 from both sides of the inequality.
$3x + 1 - 1 \leq 10 - 1$ becomes $3x \leq 9$.

Simplify.
$x \leq 3$

The mystery number is at most three.

Problem 2: The sum of two consecutive integers is at least thirteen. What could the first integer be?

First change the problem from Plain English into Math Talk.
Change "the sum of two consecutive integers" to "$x + (x + 1)$."
Change "is at least" to "$\geq$."
Change "thirteen" to "13."
In Math Talk, the problem now reads $x + (x + 1) \geq 13$.

Solve the problem. First simplify.
$x + x + 1 \geq 13$ becomes $2x + 1 \geq 13$.

Subtract 1 from both sides of the equation.
$2x + 1 - 1 \geq 13 - 1$

Simplify.
$x \geq 6$

The first of the two consecutive integers must be greater than or equal to six.

SUPER BRAIN TICKLERS

Solve for x.

1. $5(x - 2) > 6(x - 1)$

2. $3(x + 4) < 2x - 1$

3. $\frac{1}{4}x - \frac{1}{2} < \frac{1}{2}$

4. $-2(x - 3) > 0$

5. $-\frac{1}{2}(2x + 2) < 5$

(Answers are on page 79.)

BRAIN TICKLERS—THE ANSWERS
Set # 19, page 66

1. False
2. True
3. False

4. True
5. True

Set # 20, page 70

1. $x > 2$

2. $x < -2$

3. $x \geq 2$

4. $x \geq -2$

5. $x \leq -2$

6. $x < 2$

Set # 21, page 75

1. $x > \dfrac{2}{3}$
2. $x < -20$

3. $x < 4$
4. $0 < x$

5. $-1 \leq x$
6. $x \geq 12$

Super Brain Ticklers, page 78

1. $x < -4$
2. $x < -13$

3. $x < 4$
4. $x < 3$

5. $x > -6$

Chapter 5

Graphing Linear Equations and Inequalities

Graphing is a way to illustrate the solutions to various equations and inequalities. Before you can learn how to graph equations and inequalities, you need to learn how to plot points and graph a line.

Imagine two number lines that intersect each other. One of these lines is horizontal and the other line is vertical. The horizontal line is called the *x-axis*. The vertical line is called the *y-axis*. The point where the lines intersect is called the *origin*. The origin is the point (0, 0). Each of the two number lines has numbers on it.

The numbers to the right of the origin are positive. The numbers to the left of the origin are negative. Look at the *y*-axis. It is like a number line that is standing straight up. The numbers on the top half of the number line are positive. The numbers on the bottom half are negative.

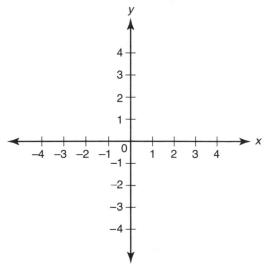

Graphing Points

You can graph points on this coordinate axis system. Each point to be graphed is written as two numbers, such as (3, 2). The first number is the *x*-value. The second number is the *y*-value. The first number tells you how far to the left or the right of the origin the point is. The second number tells you how far above or below the origin the point is.

To graph a point, follow these four *painless* steps.

Step 1: Put your pencil at the origin.

Step 2: Start with the *x*-term. It is the first term in the parentheses.

Move your pencil *x* spaces to the left if the *x*-term is negative.
Move your pencil *x* spaces to the right if the *x*-term is positive.
Keep your pencil at this point.

Step 3: Look at the *y*-term. It is the second term in the parentheses.

Move your pencil *y* spaces down if the *y*-term is negative.
Move your pencil *y* spaces up if the *y*-term is positive.

Step 4: Mark this point.

Graph (4, 2).
To graph the point (4, 2), put your pencil at the origin.
Move your pencil four spaces to the right.
Move your pencil two spaces up.
Mark this point. This is the point (4, 2).

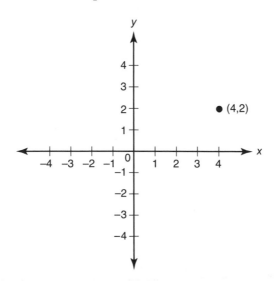

Graph (−2, 0).
To graph point (−2, 0), put your pencil at the origin.
Move your pencil two spaces to the left.
Do not move your pencil up or down, since the *y*-value is 0.
The point (−2, 0) lies exactly on the *x*-axis.

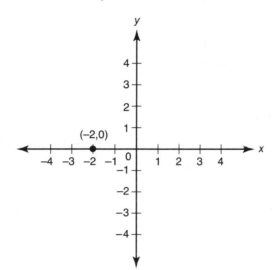

 BRAIN TICKLERS   Set # 22

For each of the following, graph the given points.

1. (1, 4)

2. (3, −1)

3. (−2, 6)

4. (−4, −2)

5. (0, 3)

6. (4, 0)

7. (0, 0)

(Answers are on page 107.)

Graphing Lines by Plotting Points

Now that you know how to graph individual points, you can graph a linear equation. The graph of a linear equation is a straight line. To graph a linear equation, find and plot three points that make the equation true. Connect these points into a straight line.

To graph a linear equation, follow these four *painless* steps.

Step 1: Solve the equation for y.

Step 2: Find three points that make the equation true.

Step 3: Graph the three points.

Step 4: Connect the three points into a straight line. Make sure to extend the line with arrows to show that it goes on forever.

Graph $x - y + 1 = 0$.
To graph the linear equation $x - y + 1 = 0$, follow the four painless steps.

Step 1: Solve the equation for y.

$$x - y + 1 = 0$$

Add y to both sides of the equation.

$$x - y + 1 + y = 0 + y$$

Simplify by combining like terms.

$$x + 1 = y$$

Step 2: Find three points that make the equation true.

Pick a number for x and figure out the corresponding y-value by substituting the number you picked for x and solving the equation.

If $x = 0, y = 1$. The point $(0, 1)$ makes the equation $y = x + 1$ true.

If $x = 1, y = 2$. The point $(1, 2)$ also makes this equation true.

If $x = 2, y = 3$. The point $(2, 3)$ also makes this equation true.

Step 3: Graph the three points.

Graph $(0, 1)$, $(1, 2)$, and $(2, 3)$.

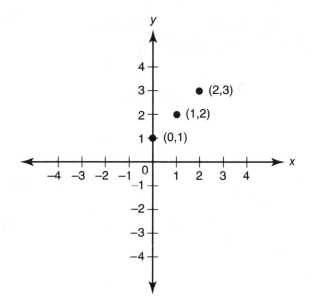

Step 4: Connect and extend the three points to make a straight line.

This is the graph of the equation $x - y + 1 = 0$.

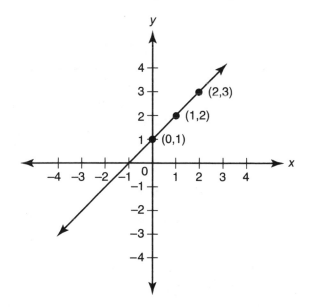

Graph $x - y = 0$.

To graph the linear equation $x - y = 0$, follow the four painless steps.

Step 1: Solve the equation $x - y = 0$ for y.

Add y to both sides of the equation.

$x - y + y = 0 + y$

Simplify by combining like terms.

$y = x$

Step 2: Find three points that make the equation true.

If $x = 0, y = 0$. The point $(0, 0)$ makes the equation $x - y = 0$ true.

If $x = 1, y = 1$. The point $(1, 1)$ makes this equation true.

If $x = 4, y = 4$. The point $(4, 4)$ makes this equation true.

Step 3: Graph the three points.

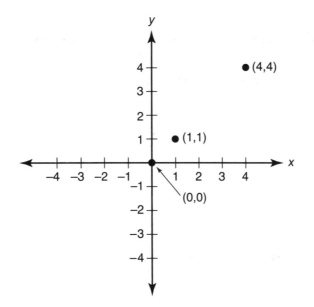

Step 4: Connect and extend the three points to make a straight line.

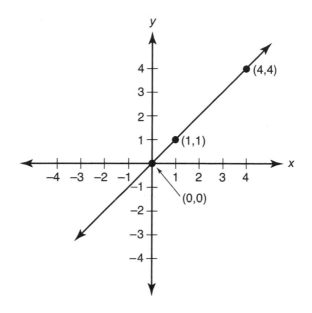

You have graphed the equation $y = x$.

Graphing Horizontal and Vertical Lines

Horizontal and vertical lines are exceptions to the graphing rule. A horizontal line does not have an x-term. It is written in the form $y = $ some number. The lines for $y = 2$, $y = 0$, and $y = -1$ are all horizontal lines.

A vertical line does not have a y-term. It is written in the form $x = $ some number. The lines for $x = 3$, $x = 0$, and $x = -\frac{1}{2}$ are all vertical lines.

Watch as these horizontal and vertical lines are graphed.

Graph $y = 3$.

If x is 0, $y = 3$. If x is 1, $y = 3$. If x is -1, $y = 3$. No matter what x equals, $y = 3$.

If x equals 987,654,321, y will still be 3.

Look at the graph. It is a horizontal line that intersects the y-axis at 3.

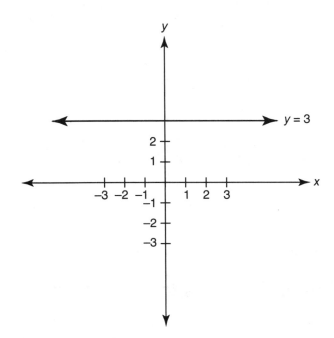

Graph $x = -2$.
If $y = 0$, $x = -2$. If y is 1, $x = -2$; and if y is 5, $x = -2$. No matter what y equals, $x = -2$.
In fact, if y were 10,000,000, x would still be equal to -2.
Look at the graph. It is a vertical line that intersects the x-axis at -2.

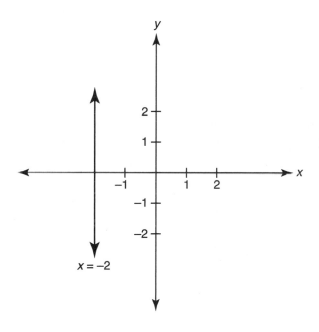

PAINLESS TIP

Any line in the form $y = 5$ intersects the y-axis. It is not parallel to the y-axis.

Any line in the form $x = 5$ intersects the x-axis. It is not parallel to the x-axis.

BRAIN TICKLERS Set # 23

Graph the following horizontal and vertical lines on the same graph.

1. $x = 2$
2. $x = -2$
3. $y = 2$
4. $y = -2$

(Answers are on page 108.)

Rate of Change

The **rate of change**, or slope, of a line is a measure of the incline of the line. The rate of change of a line is a rational number. This number indicates both the *direction* of the line and the *steepness* of the line. You can tell the direction of a line by looking at the sign of the rate of change. A positive rate of change indicates that the line goes uphill if you are moving from left to right. A negative rate of change indicates that the line goes downhill if you are moving from left to right. Note the directions of the lines in the sketches.

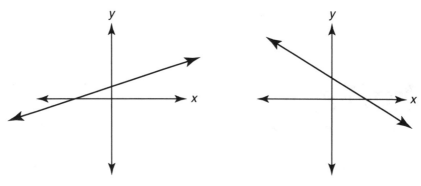

Positive Rate of Change   Negative Rate of Change

You can tell the steepness of a line by looking at the absolute value of the rate of change.

A line with a rate of change of 3 is steeper than a line with a rate of change of 1. Note the steepness of each line in the sketches.

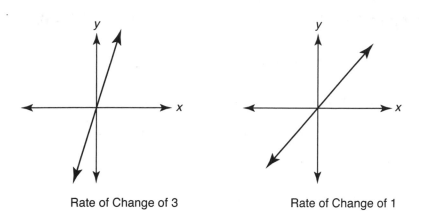

Rate of Change of 3   Rate of Change of 1

A line with a rate of change of -3 is steeper than a line with a rate of change of -1. Remember: $|-3| > |-1|$.

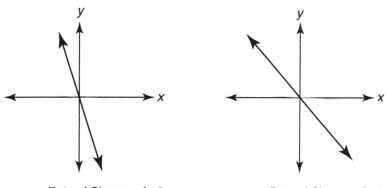

Rate of Change of -3 Rate of Change of -1

There are two common methods of finding the rate of change of a line.

1. Putting the equation in rate of change-intercept form.

2. Using the point-point method.

Method 1: Rate of Change-Intercept Method

Finding the rate of change of an equation is *painless*. There's only one simple step.

Solve the equation for y to put the equation in rate of change-intercept form. The variable in front of x is the rate of change.

Example: Find the rate of change of the line $4x + y - 2 = 0$.

Solve the equation for y.
The result is $y = -4x + 2$.
The number in front of x is the rate of change.
The rate of change of this equation is -4.

Example: Find the rate of change of $-6x + 3y = 18$.

Solve the equation for y.
The result is $y = 2x + 6$.
The number in front of x is the rate of change.
The rate of change of this equation is 2.

Example: Find the rate of change of $2y = -x$.

Solve the equation for y.
Divide both sides by 2.
The result is $y = -\dfrac{1}{2} x$.
The number in front of x is the rate of change.

The rate of change of this equation is $-\dfrac{1}{2}$. Notice that the rate of change can be a fraction.

BRAIN TICKLERS   Set # 24

Change each of the following equations to rate of change-intercept form, and find the rate of change.

1. $5x - y + 2 = 0$
2. $4x + 2y = 0$
3. $x + y = 1$
4. $-6 = 6x + 6y$

(Answers are on page 108.)

Method 2: Point-Point Method

To find the rate of change of a line using the point-point method, just complete these four *painless* steps.

Step 1: Find two points on the line.

Step 2: Subtract the first y-coordinate (y_1) from the second y-coordinate (y_2) to find the change in y.

Step 3: Subtract the first x-coordinate (x_1) from the second x-coordinate (x_2) to find the change in x.

Step 4: Divide the change in y (Step 2) by the change in x (Step 3). The answer is the rate of change of the line.

$$\text{Rate of Change} = \frac{y_2 - y_1}{x_2 - x_1}$$

Example: Find the rate of change of the line through the points $(1, 4)$ and $(3, 6)$.

Step 1: Find two points on the line. The two points $(1, 4)$ and $(3, 6)$ are given.

Step 2: Subtract the first y-coordinate (y_1) from the second y-coordinate (y_2) to find the change in y.

$$6 - 4 = 2$$

Step 3: Subtract the first x-coordinate (x_1) from the second x-coordinate (x_2) to find the change in x.

$$3 - 1 = 2$$

Step 4: Divide the change in y (Step 2) by the change in x (Step 3). The answer is the rate of change of the line.

$$\frac{2}{2} = 1$$

The rate of change is 1.

ORDER MATTERS

When finding the rate of change of a line using the point-point method, be sure to keep the same order when going from y's to x's.

Find the rate of change of the line through the two points: (1, 5) and (3, 9).

The rate of change of the line is determined by computing

$$\frac{y_2 - y_1}{x_2 - x_1}$$

The rate of change of this line is computed by subtracting the coordinates of the first point from the coordinates of the second point.

$$\frac{(9-5)}{(3-1)} = \frac{4}{2} = 2$$

The rate of change of this line is 2.

The rate of change of this line could also be computed by subtracting the coordinates of the second point from the coordinates of the first point.

$$\frac{(5-9)}{(1-3)} = \frac{-4}{-2} = 2$$

The answer is still 2.

But if you do not subtract the x's and y's in the same order, the answer will be incorrect. Watch.

$$\frac{(9-5)}{(1-3)} = \frac{4}{-2} = -2$$

The rate of change of the line through the points (1, 5) and (3, 9) is not −2.

BRAIN TICKLERS   Set # 25

For each of the following, find the rate of change of the line through the given points.

1. (5, 0) and (0, 2)

2. (0, 0) and (5, 5)

3. (−1, −4) and (−2, −4)

4. (3, 5) and (3, 1)

(Answers are on page 109.)

Finding the Equation of a Line

If you know the rate of change of a line and a point on the line, you can find the equation of the line. Just follow these three *painless* steps to find the equation of a line.

Step 1: Substitute the rate of change of the line for the variable m in the equation $y = mx + b$, and substitute the coordinates of the point on the line for the variables x and y in the same equation.

Step 2: Solve for b.

Step 3: Substitute m and b into the equation $y = mx + b$ to find the equation of the line.

Example: Find the equation of the line with rate of change -2 and the point $(-1, 1)$.

Step 1: Substitute the rate of change of the line for m in the equation $y = mx + b$, and substitute the coordinates of the point on the line for x and y in the same equation.

The equation becomes $1 = (-2)(-1) + b$.

Step 2: Solve for b.

$$1 = (-2)(-1) + b$$
$$1 = 2 + b$$
$$b = -1$$

Step 3: Substitute m and b into the equation $y = mx + b$ to find the equation of the line.

$$y = -2x - 1$$

If you know any two points on a line, you can also find the equation of the line. Just follow these four *painless* steps to find the equation of a line.

Step 1: Use the points to find the rate of change of the line. Divide the change in y by the change in x.

Step 2: Substitute the rate of change for m in the equation $y = mx + b$.

Step 3: Substitute m and one pair of coordinates for x and y in the equation, and solve for b.

Step 4: Substitute the values for m and b in the equation $y = mx + b$ to find the equation of the line.

Example: Find the equation of the line that passes through the points $(3, 3)$ and $(4, 6)$.

Step 1: Find the rate of change of the line. Divide the change in y by the change in x.

$$\frac{6-3}{4-3} = \frac{3}{1} = 3$$

The rate of change is 3.

Step 2: Substitute the rate of change for m in the equation $y = mx + b$.
$$y = 3x + b$$

Step 3: Substitute m and one pair of coordinates for x and y in the equation, and solve for b.

$$y = mx + b$$
$$3 = 3(3) + b$$
$$3 = 9 + b$$
$$3 - 9 = b$$
$$-6 = b$$

Step 4: Substitute the values for m and b in the equation $y = mx + b$ to find the equation of the line.
$$y = 3x - 6$$

BRAIN TICKLERS   Set # 26

a. Find the equation of each line.

1. Rate of change 3 and point (4, −4)

2. Rate of change $\frac{1}{4}$ and point (0, 0)

3. Rate of change −1 and point (5, 2)

b. Find the equation of the line through each pair of points.

4. (1, 1) and (3, 0)

5. (0, 4) and (2, 0)

6. (−1, 6) and (5, −2)

(Answers are on page 109.)

Graphing Using Rate of Change-Intercept Method

The easiest way to graph a line is to use the rate of change-intercept method. The expression *rate of change-intercept* refers to the form of the equation. An equation in rate of change-intercept form is written in terms of a single *y*.

Equations in rate of change-intercept form have the form $y = mx + b$, where

   y is a variable,
   m is the rate of change,
   b stands for the point where the line intercepts the *y*-axis.

Follow these five *painless* steps to graph an equation using the rate of change-intercept method.

Step 1: Put the equation in rate of change-intercept form. Express the equation of the line in terms of a single *y*. The equation should have the form $y = mx + b$.

Step 2: The number without any variable after it (the *b*-term) is the *y*-intercept. The *y*-intercept is the point where the line crosses the *y*-axis. Make a mark on the *y*-axis at the *y*-intercept. If the equation has no *b*-term, the *y*-intercept is 0.

Step 3: The number before the x-term is the rate of change. In order to graph the line, the rate of change must be written as a fraction. If the rate of change is a fraction, leave it as it is. If it is a whole number, place it over the number 1.

Step 4: Start at the y-intercept, and move your pencil up the y-axis the number of spaces in the numerator of the rate of change. Next move your pencil to the right or left the number of places in the denominator of the rate of change. If the fraction is negative, move your pencil to the left. If the fraction is positive, move your pencil to the right.

Step 5: Connect the point on the y-axis to the second point.

Example: Graph the line $2y - 3x = 4$.

Step 1: Put the equation in rate of change-intercept form.
$$y = \frac{3}{2}x + 2$$

Step 2: The number without any variable after it (the b-term) is the y-intercept. Make a mark on the y-axis at the y-intercept. In this equation, the y-intercept is 2.

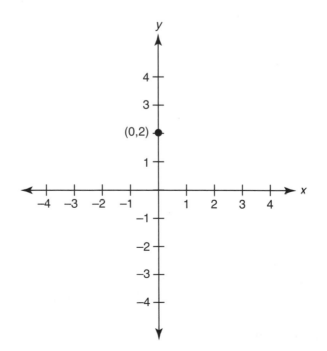

Step 3: The number before the x-term is the rate of change.

$\frac{3}{2}$ is the rate of change.

Step 4: Starting at the y-intercept, move your pencil up the y-axis the number of spaces in the numerator of the rate of change. Next, move your pencil to the left or right the number of spaces in the denominator. Mark the point.

Since the rate of change is $\frac{3}{2}$, move your pencil up

three spaces and to the right two spaces. The point is $(2, 5)$. Mark it.

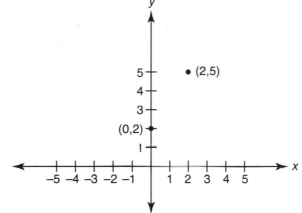

Step 5: Connect the point on the y-axis to the point you just marked.

Connect $(0, 2)$ to $(2, 5)$.
This is the graph of the line $2y - 3x = 4$.

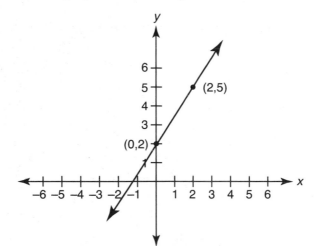

REMINDER

An equation in rate of change-intercept form has the form $y = mx + b$, where m is the rate of change and b is the point where the line intercepts the y-axis.

BRAIN TICKLERS   Set # 27

What are the rate of change and the y-intercept of each of these lines? Graph the lines using the rate of change-intercept method.

1. $y = x + 1$

2. $y = x - 1$

3. $3x + 3y = 12$

4. $2x + 6y = 16$

(Answers are on page 109.)

Graphing Inequalities

How do you graph an inequality such as $2x - 4 > y$ or $x - 6y \leq 0$? It's not as hard as you may think. If you can graph a straight line, you can graph an inequality.

Just follow these three *painless* steps.

Step 1: Graph the inequality as if it were an equation. You can plot points or use the rate of change-intercept method.

Step 2: If the inequality reads "$\leq$" or "$\geq$," leave the line solid since it is included in the graph. If the inequality reads "$<$" or "$>$," use your eraser to make the line dashed. The line is not included in the solution.

Step 3: Pick a test point not on the line, usually $(0, 0)$ or $(1, 1)$. Substitute this point into the inequality to test whether this point makes the inequality true or false. If it is true, shade the side of the line that contains the point. If it is false, shade the graph on the side of the line that does not contain the point.

$2x - 4 > y$

Step 1: Graph the inequality as if it were an equation. First, rewrite $2x - 4 > y$ as an equation.
   $2x - 4 > y$ becomes $2x - 4 = y$.
   Graph using the rate of change-intercept method.
      The rate of change of the equation $2x - 4 = y$ is 2.
      The y-intercept is -4.
      Now that you know the rate of change and y-intercept, you can graph the line.

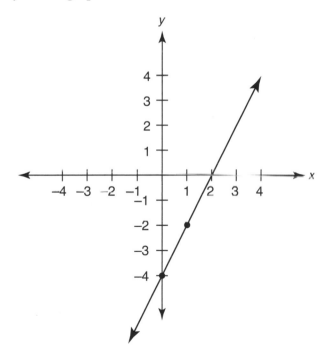

Step 2: If the inequality reads "<" or ">," use your eraser to make the line dashed.

Make the line of the graph $2x - 4 > y$ dashed.

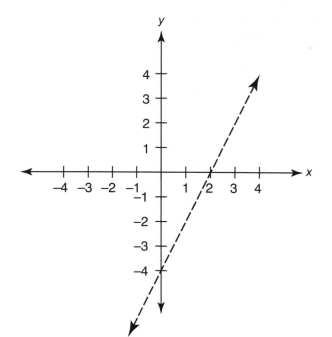

Step 3: Pick a test point not on the line, usually $(0, 0)$ or $(1, 1)$. Substitute this point into the inequality to test whether this point makes the inequality true or false. If it is true, shade the side of the line that contains the point. If it is false, shade the graph on the side of the line that does not contain the point.

Pick the point $(0, 0)$ and plug it into the inequality $2x - 4 > y$.
$2(0) - 4 > 0$
Simplify: $-4 > 0$.

This statement is false.

Since it is false, shade the side of the line that does not contain the point $(0, 0)$.

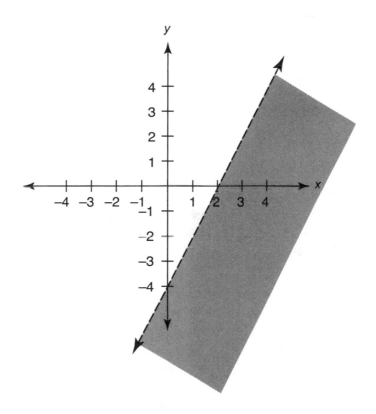

Now let's graph another inequality.

$x - 6y \leq 12$

Step 1: Graph the inequality as if it were an equation.

$x - 6y \leq 12$ becomes $x - 6y = 12$.

Graph using two points.

Set $x = 0$, and solve for y.

If $x = 0$, then $0 - 6y = 12$.

Solve $-6y = 12$ for y: $y = -2$.

The first point is $(0, -2)$.

Next set $y = 0$, and solve for x.

If $y = 0$, then $x - 6(0) = 12$.

Solve this equation: $x = 12$.

The second point is $(12, 0)$.

Plot the two points, and connect them to graph the line.

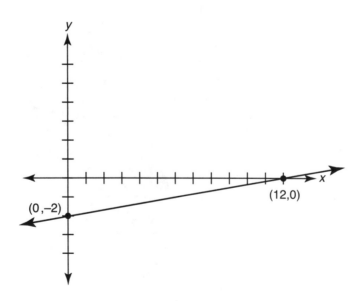

Step 2: If the inequality reads $<$ or $>$, use your eraser to make the line dashed. The graph of the inequality $x - 6y \leq 12$ should be a solid line, because $x - 6y$ is less than *or equal to* 12.

Step 3: Pick a test point not on the line, usually $(0, 0)$ or $(1, 1)$. Substitute this point into the inequality to test whether this point makes the inequality true or false. If it is true, shade the side of the line that contains the point. If it is false, shade the graph on the side of the line that does not contain the point.

Pick the point $(0, 0)$ and plug it into the inequality $x - 6y \leq 12$.
$0 - 6(0) \leq 12$
Simplify: $0 \leq 12$.

This statement is true.

Since it is true, shade the graph on the side of the line that contains $(0, 0)$.

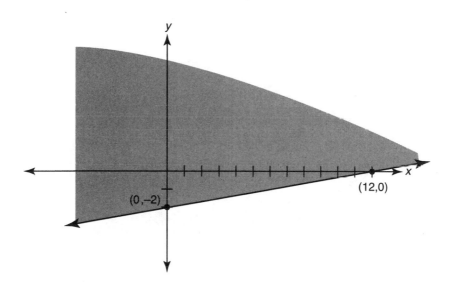

BRAIN TICKLERS   Set # 28

Graph each of the following inequalities.

1. $2x > y + 6$

2. $y < 2x - 1$

3. $y \leq x$

4. $x > 2$

(Answers are on page 111.)

BRAIN TICKLERS—THE ANSWERS
Set # 22, page 83

1.

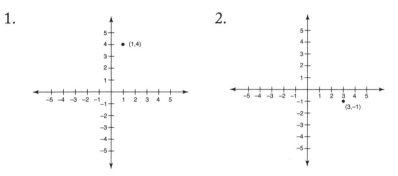

2.

3.

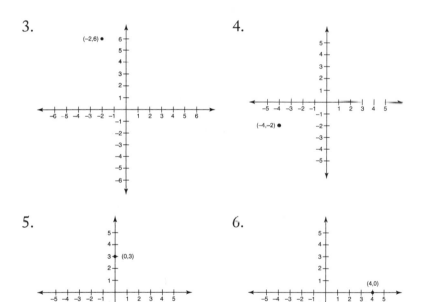

4.

5.

6.

7.

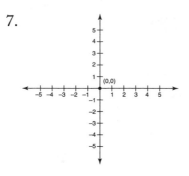

Set # 23, page 90

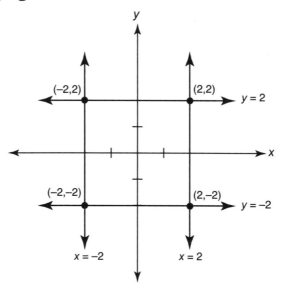

Set # 24, page 92

1. $y = 5x + 2$, rate of change $= 5$

2. $y = -2x$, rate of change $= -2$

3. $y = 1 - x$, rate of change $= -1$

4. $y = -x - 1$, rate of change $= -1$

Set # 25, page 95

1. $-\dfrac{2}{5}$　　2. 1　　3. 0　　4. Undefined

Set # 26, page 97

a.　1. $y = 3x - 16$ 　　　　　　*b.*　4. $y = -\dfrac{1}{2}x + \dfrac{3}{2}$

　　2. $y = \dfrac{1}{4}x$ 　　　　　　　　　5. $y = -2x + 4$

　　3. $y = -x + 7$ 　　　　　　　　6. $y = -\dfrac{4}{3}x + \dfrac{14}{3}$

Set # 27, page 100

1. Rate of change 1, y-intercept 1

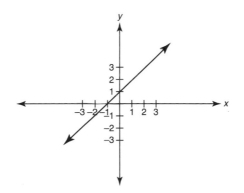

2. Rate of change 1, y-intercept -1

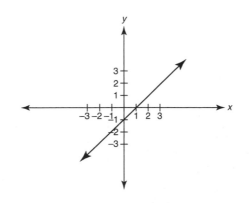

3. Rate of change -1, y-intercept 4

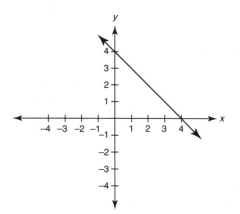

4. Rate of change $-\dfrac{1}{3}$, y-intercept $2\dfrac{2}{3}$

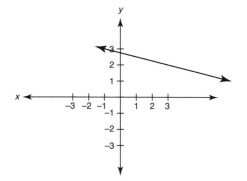

Set # 28, page 106

1. $2x > y + 6$

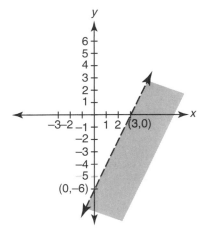

2. $y < 2x - 1$

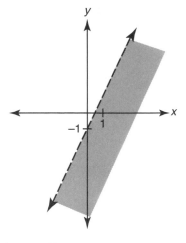

3. $y \leq x$

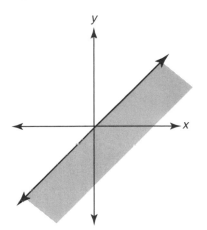

4. $x > 2$

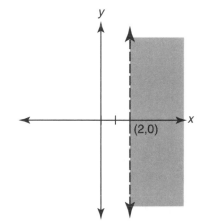

Chapter 6

Solving Systems of Equations and Inequalities

Some equations have more than one variable. In this chapter, you will learn how to solve equations with two variables. Here are examples of equations with two variables:

$$3x + 2y = 7$$
$$4x - y = -5$$
$$x - y = 0$$

It's impossible to solve a single equation with two variables and get a single solution. For example, for the equation $x - y = 0$, there are many possible answers. If $x = 5$ and $y = 5$, then $x - y = 0$. If $x = 100$ and $y = 100$, then $x - y = 0$. In fact, any time $x = y$, then $x - y = 0$.

But if you have two equations, each with two variables, such as $x + 3 = y$ and $x + y = 5$, only one value of x and only one value of y will make both of these equations true. Pairs of equations with the same two variables are called *systems of linear equations*. Here are three examples of systems of linear equations:

$$x + y = 5 \text{ and } x - y = 5$$
$$2x - 3y = 7 \text{ and } 3x - 4y = 12$$
$$6x = 3y \text{ and } x + y = 3$$

In this chapter, you will learn how to solve systems of linear equations like these using three different techniques: addition or subtraction, substitution, and graphing.

Solving Systems of Linear Equations by Addition or Subtraction

In order to solve a system of linear equations, first determine the relationship between the pair of equations. In a system of linear equations, there are three possibilities for the relationship between the pairs of equations.

Relationship 1: In the pair of equations, the coefficient of one of the x-terms is the opposite of the coefficient of the x-term in the other equation. This may sound complicated, but it's *painless*.
In this pair of linear equations, the coefficients in front of the x-terms are the opposite of each other; -6 is the opposite of 6.

$$-6x + 2y = 0$$
$$6x + 4y = 3$$

Or, in the pair of equations, the coefficient of one of the y-terms is the opposite of the coefficient of the y-term in the other equation.

$$x + 3y = 7$$
$$x - 3y = -2$$

Notice that in this pair of equations the coefficients in front of the y-terms are 3 and -3; 3 and -3 are the opposites of each other.

When you find a pair of equations of this type, it is easy to solve the system. Just use the following steps.

Step 1: Add the two equations.

Step 2: Solve the resulting equation.

Step 3: Substitute the answer in one of the original equations to solve for the other variable.

Step 4: Check the answer.

Now watch as the following system of two equations is solved using addition. Notice that the coefficients in front of the y-terms are the opposites of each other.

Solve: $3x - 2y = 5$
$3x + 2y = 13$

Step 1: Add the two equations.

Watch what happens when you add these two equations.
$$3x - 2y = 5$$
$$\underline{3x + 2y = 13}$$
$$6x\quad\ \ = 18$$

Step 2: Solve the resulting equation.

Solve $6x = 18$. Divide both sides of this equation by 6.

$x = 3$

Step 3: Substitute this answer in one of the original equations to solve for the other variable.

Substitute $x = 3$ in the equation $3x - 2y = 5$.
The new equation is $3(3) - 2y = 5$.
Solve.
$y = 2$

Step 4: Check the answer.

The answer is $x = 3$ and $y = 2$.
Check this answer by substituting these values for x and y in the original two equations, $3x - 2y = 5$ and $3x + 2y = 13$.
Substitute $x = 3$ and $y = 2$ in $3x - 2y = 5$.
The resulting equation is $3(3) - 2(2) = 5$.

$9 - 4 = 5$

This is a true sentence.

Now substitute $x = 3$ and $y = 2$ in the second equation, $3x + 2y = 13$.
The resulting equation is $3(3) + 2(2) = 13$.

$9 + 4 = 13$

This is a true sentence.
The pair $x = 3$ and $y = 2$ makes $3x - 2y = 5$ *and* $3x + 2y = 13$ true.

Now watch as two more equations are solved using addition. Notice that the coefficients in front of the x-variable are 1 and -1.

Solve: $x + 4y = 17$
          $-x - 2y = -9$

Step 1: Add the two equations.

$$x + 4y = 17$$
$$\underline{-x - 2y = -9}$$
$$2y = 8$$

Step 2: Solve the resulting equation.

Solve $2y = 8$.

$y = 4$

Step 3: Substitute this answer in one of the original equations to solve for the other variable.

Substitute $y = 4$ in the equation $x + 4y = 17$.

$x + 4(4) = 17$

Solve.

$x = 1$

Step 4: Check the answer.

The answer is $x = 1$ and $y = 4$.

Substitute this answer in the original equations, $x + 4y = 17$ and $-x - 2y = -9$.

Substitute $x = 1$ and $y = 4$ into $x + 4y = 17$.

$(1) + 4(4) = 17$

This is a true sentence.

Substitute $x = 1$ and $y = 4$ in the other original equation, $-x - 2y = -9$.

$-(1) - 2(4) = -9$

$-1 - 8 = -9$

This is also a true sentence, so $x = 1$ and $y = 4$ make both equations true.

BRAIN TICKLERS   Set # 29

Solve the following systems of equations using addition.

1. $x - y = 4$

   $x + y = 8$

2. $3x + y = 0$

   $-3x + y = -6$

3. $2x + \frac{1}{4} y = -1$

   $x - \frac{1}{4} y = -2$

(Answers are on page 136.)

Relationship 2: Sometimes two equations have the same coefficient in front of one of the variables. Following are two examples of equations with the same coefficients.

$$4x - y = 7$$
$$4x + 2y = 10$$

The coefficients in front of the x-variable are both 4.
The coefficients in front of the y-variable in these equations are both $-\frac{1}{2}$.

$$3x - \frac{1}{2}y = 6$$

$$-2x - \frac{1}{2}y = -3$$

To solve equations with the same coefficient, follow these four *painless* steps.

Solve: $\frac{1}{4}x + 3y = 6$

$\qquad \frac{1}{4}x + y = 4$

Step 1: Subtract one equation from the other.

In order to subtract $\frac{1}{4}x + y = 4$ from $\frac{1}{4}x + 3y = 6$, you first have to distribute the negative sign in front of the equation: $-\left(\frac{1}{4}x + y = 4\right)$. Change the negative sign to a (-1) so the expression becomes $(-1)\left(\frac{1}{4}x + y = 4\right)$. Multiply -1 by every number inside the parentheses: $(-1)\left(\frac{1}{4}x\right) + (-1)(y) = (-1)(4)$, which equals $-\frac{1}{4}x - y = -4$.

Now add the two equations.

$$\frac{1}{4}x + 3y = 6$$
$$-\frac{1}{4}x - y = -4$$
$$\overline{}$$
$$2y = 2$$

Step 2: Solve the resulting equation.

Solve $2y = 2$.

$y = 1$

Step 3: Substitute this answer in one of the original equations. Solve for the other variable.

Substitute $y = 1$ in $\frac{1}{4}x + y = 4$.

$$\frac{1}{4}x + 1 = 4$$

Simplify.

$$\frac{1}{4}x = 3$$

Simplify.

$$x = 12$$

Step 4: Check your answer.

Substitute $x = 12$ and $y = 1$ into the original equations,

$\frac{1}{4}x + 3y = 6$ and

$\frac{1}{4}x + y = 4$.

Equation 1: $\frac{1}{4}(12) + 3(1) = 6$

$3 + 3 = 6$

$6 = 6$

Equation 2: $\frac{1}{4}(12) + (1) = 4$

$3 + 1 = 4$

$4 = 4$

Here is an example of another pair of equations that have the same coefficient. Notice that each of these equations has a -2 in front of the y-variable.

Solve: $4x - 2y = 4$

$-2x - 2y = 10$

To solve these equations, follow the four-step *painless* method.

Step 1: Subtract the second equation from the first equation.

$$\begin{array}{r} 4x - 2y = 4 \\ -(-2x - 2y = 10) \\ \hline 6x \quad\quad = -6 \end{array}$$

Step 2: Solve the resulting equation.

Solve $6x = -6$.

Divide both sides of the equation by 6.

$x = -1$

Step 3: Substitute this answer in one of the original equations to solve for the other variable.

Substitute $x = -1$ in $4x - 2y = 4$

$4(-1) - 2y = 4$

Solve.

$-4 - 2y = 4$

Simplify.

$y = -4$

Step 4: Check your answer.

Substitute $x = -1$ and $y = -4$ into $4x - 2y = 4$ and $-2x - 2y = 10$.

Equation 1: $4(-1) - 2(-4) = 4$
$-4 - (-8) = 4$
$4 = 4$

Equation 2: $-2(-1) - 2(-4) = 10$
$2 - (-8) = 10$
$10 = 10$

BRAIN TICKLERS   Set # 30

Solve the following systems of equations using subtraction.

1. $3x - 2y = 7$

   $6x - 2y = 4$

2. $\frac{1}{2}x + 2y = 3$

   $\frac{1}{2}x - 5y = 10$

3. $3x + 6y = 9$

   $2x + 6y = 8$

4. $7x - \frac{2}{3}y = 12$

   $x - \frac{2}{3}y = 0$

(Answers are on page 136.)

Relationship 3: Sometimes the coefficients of the two equations have no relationship to each other.

$$2x - 5y = 2$$
$$-5x + 3y = 4$$

Adding these two equations will not help solve them.
Subtracting the second equation from the first will not help solve them either.
In order to solve equations of this type, follow these six *painless* steps.

Step 1: Multiply the first equation by the coefficient in front of x in the second equation.

Step 2: Multiply the second equation by the coefficient in front of x in the first equation.

Step 3: Add or subtract the two new equations.

Step 4: Solve the resulting equation.

Step 5: Substitute this answer in the original equation to solve for the other variable.

Step 6: Check your answer.

Now watch as these two equations are solved.

Solve: $2x - 4y = 0$
$-5x + 2y = 4$

Step 1: Multiply the first equation by the coefficient in front of x in the second equation.

Negative five is the coefficient in front of x in the second equation.
Multiply the first equation by -5.

$$(-5)(2x - 4y = 0)$$
$$-5(2x) - 5(-4y) = -5(0)$$
$$-10x + 20y = 0$$

Step 2: Multiply the second equation by the coefficient in front of x in the first equation.

Two is the coefficient in front of x in the first equation. Multiply the second equation by 2.

$$(2)(-5x + 2y = 4)$$
$$2(-5x) + 2(2y) = 2(4)$$
$$-10x + 4y = 8$$

Step 3: Add or subtract the two new equations.

Subtract the equations.

$$\begin{array}{r} -10x + 20y = 0 \\ -10x + 4y = 8 \\ \hline 16y = -8 \end{array}$$

Step 4: Solve the resulting equation.

$16y = -8$
Divide both sides of this equation by 16.
$$y = -\frac{1}{2}$$

Step 5: Substitute this answer in the original equation to solve for the other variable.

$$2x - 4\left(-\frac{1}{2}\right) = 0$$

Multiply.

$$2x + 2 = 0$$

$$x = -1$$

Step 6: Check your answer.

$$x = -1 \ \text{ and } \ y = -\frac{1}{2}$$

For practice, check these answers by substituting them back in both of the original equations.

Here is another solution of a system of two equations.

Solve: $3x - 2y = 9$
$\qquad -x + 3y = 4$

Adding or subtracting these two equations will not help solve them. Use the six-step process for solving equations with no relationship.

Step 1: Multiply the first equation by the coefficient in front of x in the second equation.

Negative 1 is the coefficient in front of x in the second equation. Multiply the first equation by -1.
$(-1)(3x - 2y = 9)$ is $-3x + 2y = -9$

Step 2: Multiply the second equation by the coefficient in front of x of the first equation.

Three is the coefficient in front of the first equation. Multiply the second equation by 3.
$(3)(-x + 3y = 4)$ is $-3x + 9y = 12$

Step 3: Add or subtract these two new equations.

Subtract the equations.
$$\begin{array}{r} -3x + 2y = -9 \\ -3x + 9y = 12 \\ \hline -7y = -21 \end{array}$$

Step 4: Solve the resulting equation.

$-7y = -21$
$y = 3$

Step 5: Substitute the value of y in the original equation to solve for x.

Substitute 3 for y in the equation $3x - 2y = 9$.
$3x - 2(3) = 9$

Solve.
$x = 5$

Step 6: Check. Substitute the values of $x = 5$ and $y = 3$ in each of the original equations. If each of the results is a true sentence, the equation is correct.

Check this solution on your own.

BRAIN TICKLERS   Set # 31

For each of the following, multiply the first equation by the coefficient of the x-term of the second equation. Multiply the second equation by the coefficient of the x-term of the first equation. Next, add or subtract to solve the system of equations.

1. $3x + 2y = 12$

    $x - y = 10$

2. $2x - y = 3$

    $-4x + y = 6$

3. $-5x + y = 8$

    $-2x + 2y = 4$

4. $\frac{1}{2}x - 2y = 6$

    $4x + 2y = 12$

(Answers are on page 136.)

Solving Systems of Linear Equations by Substitution

Substitution is another way to solve a system of linear equations. Here is how you solve two equations with two variables using substitution.

Follow these five *painless* steps.

Step 1: Solve one of the equations for x. The answer will be in terms of y.

Step 2: Substitute this value for x in the other equation. Now there is one equation with one variable.

Step 3: Solve the equation for y.

Step 4: Substitute the value of y in one of the original equations to find the value of x.

Step 5: Check. Substitute the values of both x and y in both of the original equations. If each result is a true sentence, the solution is correct.

Watch as this system of two linear equations is solved using substitution.

Solve: $x - y = 3$
$2x + y = 12$

Step 1: Solve one of the equations for x. The answer will be in terms of y.

Solve $x - y = 3$ for x.
$$x = y + 3$$

Step 2: Substitute this value for x in the other equation. Substitute $(y + 3)$ into $2x + y = 12$ wherever there is an x.

$$2(y + 3) + y = 12$$

Step 3: Solve the equation for y.
$$2(y + 3) + y = 12$$

Simplify.
$$2y + 6 + y = 12$$
$$3y + 6 = 12$$

Simplify.
$$3y = 6$$

Solve.
$$y = 2$$

Step 4: Substitute the value of y in one of the original equations to find the value of x.

Substitute $y = 2$ in the equation $x - y = 3$.
$$x - 2 = 3$$

Solve.
$$x = 5$$

Step 5: Check. Substitute the values of both x and y in the original equations. If each result is a true sentence, the solution is correct.

Substitute $x = 5$ and $y = 2$ into $2x + y = 12$ to check the answer.

$$2(5) + 2 = 12$$

Compute the value of this expression.

$$10 + 2 = 12$$
$$12 = 12$$

This is a true sentence. You can also check that the values $x = 5$ and $y = 2$ make the equation $x - y = 3$ true. The solution $x = 5$ and $y = 2$ is correct.

Watch as two more linear equations are solved using substitution.

Solve: $x + 3y = 6$
$x - 3y = 0$

Step 1: Solve one of the equations for x. The answer will be in terms of y.

Solve $x - 3y = 0$ for x.
$$x = 3y$$

Step 2: Substitute this value for x in the other equation.
Substitute $x = 3y$ in the equation $x + 3y = 6$.
$$3y + 3y = 6$$
Now there is one equation with one variable.

Step 3: Solve the equation $3y + 3y = 6$ for y.
$$y = 1$$

Step 4: Substitute the value of y in one of the original equations to find the value of x.
Substitute $y = 1$ in the equation $x + 3y = 6$.
$$x + 3(1) = 6$$
$$x = 3$$

Step 5: Check. Substitute the values of both x and y in both of the original equations. If each result is a true sentence, the solution is correct.

$$x = 3 \text{ and } y = 1$$

Substitute these numbers in the equation $x - 3y = 0$.

$$3 - 3(1) = 0$$
$$3 - 3 = 0$$

This is a true sentence. You can also check that the values $x = 3$ and $y = 1$ make the equation $x + 3y = 6$ true. The solution $x = 3$ and $y = 1$ is correct.

BRAIN TICKLERS   Set # 32

Solve the following systems of equations using substitution.

1. $x + y = 7$

   $x - y = 1$

2. $2x + 5y = 7$

   $x + y = 2$

3. $2x + y = 0$

   $x + y = -2$

4. $2x + y = 3$

   $4x + 3y = 8$

(Answers are on page 136.)

Solving Systems of Linear Equations by Graphing

To solve a system of linear equations by graphing, follow these four steps.

Step 1: Graph the first equation.

Step 2: Graph the second equation on the same set of axes.

Step 3: Find the solution, which is the point where the two lines intersect.

Step 4: Check the answer. Substitute the intersection point in each of the two original equations. If each equation is true, the answer is correct.

Watch as graphing is used to solve a system of two linear equations.

Solve: $2x + y = 1$
$x - y = -1$

Step 1: Graph the first equation, $2x + y = 1$.
Solve the equation for y.

$$y = 1 - 2x$$

Now find three points by substituting 0, 1, and 2 for x.

If $x = 0, y = 1$.
If $x = 1, y = -1$.
If $x = 2, y = -3$.

Now graph these points. Connect and extend them.

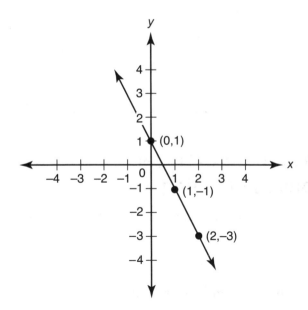

Step 2: Graph the second equation, $x - y = -1$, on the same axes. First, rewrite the equation in terms of y.

$$y = x + 1$$

Now find three points that make the equation $y = x + 1$ true.

If $x = 0, y = 1$.
If $x = 1, y = 2$.
If $x = 2, y = 3$.

Graph these points, and connect them on the same graph that you made for the first equation.

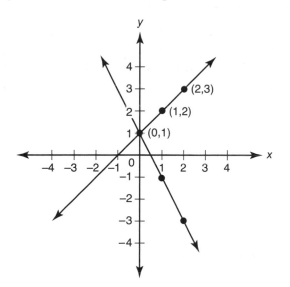

Step 3: The solution is the point where the two lines intersect, that is, $(0, 1)$.

Step 4: Check the answer. Substitute the intersection point in each of the two original equations. If both sentences are true, then the answer is correct.
Substitute the point $(0, 1)$ in the equation $2x + y = 1$.
Substitute 0 for x and 1 for y.

$$2(0) + 1 = 1$$

Compute.

$$1 = 1$$

Substitute the point $(0, 1)$ in the equation $x - y = -1$.
Substitute 0 for x and 1 for y.

$$0 - 1 = -1$$

Compute.

$$-1 = -1$$

Both sentences are true.
The solution is correct.

Solving Systems of Linear Inequalities by Graphing

Now that you know how to graph linear inequalities, you can graph two or more inequalities on the same graph to solve a system of inequalities. Where the graphs intersect (overlap) is the solution to the system of linear inequalities. Watch. The method is *painless* and fun.

Find the solution to this system of inequalities.
$y < x + 2$
$y \geq -1$
$x < 4$

First, graph the inequality $y < x + 2$.

Clue: Graph $y < x + 2$. Test $(0, 0)$. It makes the equation true. Lightly shade below the *dashed* line.

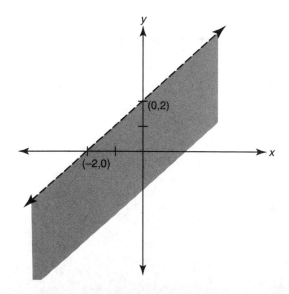

Next, on the same graph, graph the inequality $y \geq -1$.

Clue: Graph $y = -1$. The result should be a *horizontal* line. Test $(0, 0)$.

Since $(0, 0)$ makes the equation true, and $(0, 0)$ is above the dotted line, lightly shade above the *solid* line.

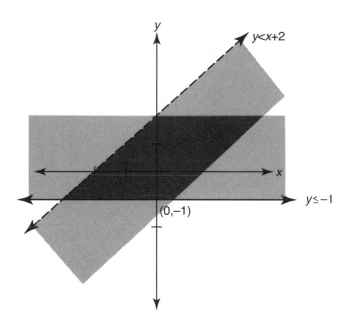

Next, on the same graph, graph the inequality $x < 4$.

Clue: Graph $x = 4$. The result should be a *vertical* line. Test $(0, 0)$.

Since $(0, 0)$ makes the equation true and $(0, 0)$ is to the left of the dotted line, lightly shade to the left of the dotted line.

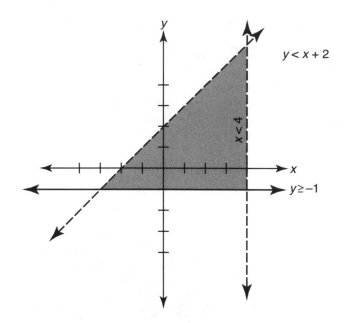

The area where the three graphs overlap is the solution set to the system of linear inequalities.

Outline and shade this area.

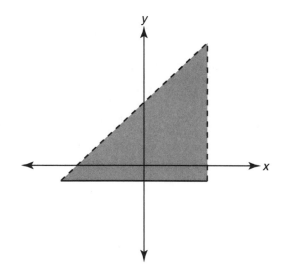

Notice that one of the sides of the triangle is a solid line and the other two are dashed lines. Why?

BRAIN TICKLERS   Set # 33

Graph the following systems of inequalities.

1. $y > x - 4$ 2. $y \geq x$ 3. $y < -5$

    $y > 0$ $x \geq 0$ $y > 2$

    $x > 1$ $y \geq 0$ $x + y < 3$

(Answers are on page 137.)

Word Problems

Watch how systems of equations can be used to solve word problems.

Problem 1: Together Keisha and Martha earned $8.
            Keisha earned $2 more than Martha.
            How much money did Keisha and Martha each earn?

To solve this word problem, change it into two equations.
Pick two letters to represent Keisha and Martha.
Let K represent Keisha and M represent Martha.
Change each given sentence into an equation.
Together Keisha and Martha earned $8.

$K + M = 8$

Keisha earned $2 more than Martha.
$K - M = 2$

Solve these two equations using addition.

$$
\begin{array}{r}
K + M = 8 \\
\underline{K - M = 2} \\
2K = 10
\end{array}
$$

Solve for K.

$$K = 5$$

Substitute 5 in the original equation $K + M = 8$, and solve for M.

$5 + M = 8$

Solve for M.

$M = 3$

The answers are $K = 5$ and $M = 3$.

To check these answers, substitute these numbers in the other equation.
Substitute $K = 5$ and $M = 3$ in the equation $K - M = 2$.
$5 - 3 = 2$
This is a true sentence.
Keisha earned \$5, and Martha earned \$3.

Problem 2: Jorge is twice as old as Sean.

Together, Jorge and Sean have lived for 18 years. How old are Jorge and Sean?

To solve this word problem, change it into two equations.
Pick two letters to represent Jorge and Sean.
Let J represent Jorge and S represent Sean.
Change each given sentence into an equation.
Jorge is twice as old as Sean.

$J = 2S$

Together, Jorge and Sean have lived for 18 years.
$J + S = 18$

Watch as these two equations are solved by substitution.

$J = 2S$
$J + S = 18$

Substitute $2S$ for J in the equation $J + S = 18$.

$2S + S = 18$

Solve.

$S = 6$

Substitute 6 for S in the original equation $J = 2S$ to solve for J.

$\quad J = 2(6)$

Multiply.

$\quad J = 12$

Substitute $S = 6$ and $J = 12$ in the other original equation, $S + J = 18$.

$\quad 6 + 12 = 18$

Simplify.

$\quad 18 = 18$

The answers are correct.

Sean is 6 years old and Jorge is 12 years old.

SUPER BRAIN TICKLERS

Solve the following systems of equations using addition.

1. $\quad x + y = 12$

$\quad\quad 2x - y = 0$

2. $-2x + 5y = 1$

$\quad\quad x - 2y = 4$

Solve the following systems of equations using substitution.

3. $\quad 3x - y = 4$

$\quad\quad -2x + y = 1$

4. $\quad x + y = 7$

$\quad\quad 3x - 2y = 4$

(Answers are on page 138.)

BRAIN TICKLERS—THE ANSWERS
Set # 29, page 117

1. $x = 6; y = 2$
2. $x = 1; y = -3$
3. $x = -1; y = 4$

Set # 30, page 120

1. $x = -1; y = -5$
2. $x = 10; y = -1$
3. $x = 1; y = 1$
4. $x = 2; y = 3$

Set # 31, page 124

1. $x = \dfrac{32}{5}, y = -\dfrac{18}{5}$

2. $x = -\dfrac{9}{2}, y = -12$

3. $x = -\dfrac{3}{2}, y = \dfrac{1}{2}$

4. $x = 4, y = -2$

Set # 32, page 127

1. $x = 4; y = 3$
2. $x = 1; y = 1$
3. $x = 2; y = -4$
4. $x = \dfrac{1}{2}; y = 2$

Set # 33, page 133

1.

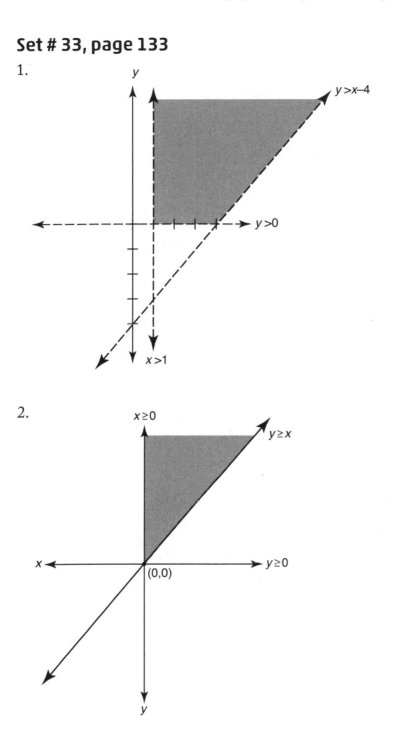

2.

3. There are no points where the equations of all three graphs intersect.

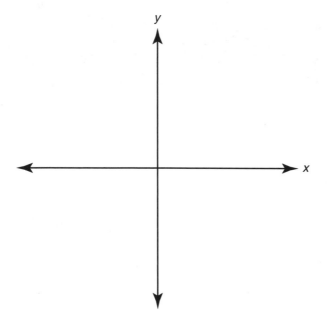

Super Brain Ticklers, page 135

1. $x = 4$; $y = 8$

2. $x = 22$; $y = 9$

3. $x = 5$; $y = 11$

4. $x = \dfrac{18}{5}$; $y = \dfrac{17}{5}$

Chapter 7

Exponents

Exponents are a shorthand way of writing repeated multiplication. The expression $2 \cdot 2 \cdot 2 \cdot 2 \cdot 2 \cdot 2$ is read as "two times two times two times two times two times two." But instead of writing and reading all that, mathematicians just say 2^6, that is, 2 to the sixth power. In the exponential expression 2^6, 2 is the *base* and 6 is the *exponent*. Think of exponents as shorthand. They are a short way of writing repeated multiplication.

1+2=3 MATH TALK!

Watch how to change these expressions from Math Talk into Plain English.

$$4^2$$
four to the second power
four squared
four times four

$$5^3$$
five to the third power
five cubed
five times five times five

$$6^4$$
six to the fourth power
six times six times six times six

$$x^5$$
x to the fifth power
x times x times x times x times x

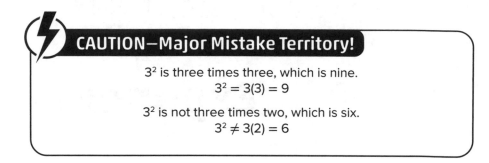

CAUTION—Major Mistake Territory!

3^2 is three times three, which is nine.
$$3^2 = 3(3) = 9$$

3^2 is not three times two, which is six.
$$3^2 \neq 3(2) = 6$$

You can compute the value of an exponential expression by multiplying.

$$2^2 = 2(2) = 4$$
$$2^3 = 2(2)(2) = 8$$
$$2^4 = 2(2)(2)(2) = 16$$
$$2^5 = 2(2)(2)(2)(2) = 32$$
$$3^2 = 3 \times 3 = 9$$
$$3^3 = 3 \times 3 \times 3 = 27$$

When you square a negative number, the answer is always positive. A negative number times a negative number is a positive number.

$$(-2)^2 = (-2)(-2) = +4$$
$$(-3)^2 = (-3)(-3) = +9$$
$$(-4)^2 = (-4)(-4) = +16$$

A negative number cubed is always a negative number.

$$(-2)^3 = (-2)(-2)(-2) = -8$$
$$(-3)^3 = (-3)(-3)(-3) = -27$$
$$(-4)^3 = (-4)(-4)(-4) = -64$$

A negative number raised to any even power is always a positive number. A negative number raised to any odd power is always a negative number.

$$(-1)^{17} = -1 \qquad\qquad (-2)^2 = 4$$
$$(-1)^{10} = +1 \qquad\qquad (-2)^3 = -8$$
$$(-1)^{25} = -1 \qquad\qquad (-2)^4 = 16$$
$$(-1)^{99} = -1 \qquad\qquad (-2)^5 = -32$$
$$(-1)^{100} = 1 \qquad\qquad (-2)^6 = 64$$
$$(-2)^7 = -128$$

Zero power

Any number to the zero power is one.

Examples:

$$6^0 = 1 \qquad\qquad 4^0 = 1 \qquad\qquad a^0 = 1$$

BRAIN TICKLERS   Set # 34

Compute the value of each of these exponential expressions.

1. 5^2

2. 2^6

3. 10^2

4. 4^3

5. 5^0

6. $(-3)^2$

7. $(-5)^3$

8. $(-4)^2$

9. $(-1)^5$

10. $(-1)^{12}$

(Answers are on page 156.)

Multiplying Exponents with Coefficients

Some exponential expressions have coefficients in front of them. The exponential expression is multiplied by its coefficient. In the expression $3(5)^2$,

> 3 is the coefficient;
> 5 is the base;
> 2 is the exponent.

In the expression $2y^3$,

>   2 is the coefficient;
> y is the base;
>   3 is the exponent.

In the expression $(5x)^2$,

>   1 is the coefficient;
> $5x$ is the base;
>   2 is the exponent.

In the expression $2x(3x)^4$,

> $2x$ is the coefficient;
> $3x$ is the base;
>   4 is the exponent.

To compute the value of an exponential expression with a coefficient, compute the exponential expression first. Next, multiply by the coefficient.

Compute the value of each of the following exponential expressions.

$5(3)^2$
First square the three.
$3^2 = 9$
Next multiply by 5.
$5(9) = 45$
The answer: $5(3)^2 = 45$

$-3(-2)^2$
First square the negative two.
$(-2)(-2) = 4$
Next multiply by -3.
$-3(4) = -12$
The answer: $-3(-2)^2 = -12$

$2(5 - 3)^2$

Do what is inside the parentheses first.

$5 - 3 = 2$

$2(2)^2$

Next square the two.

$2^2 = 4$

Finally, multiply two times the result.

$2(4) = 8$

The answer: $2(5 - 3)^2 = 8$

BRAIN TICKLERS   Set # 35

Compute the value of each of the following exponential expressions.

1. $3(5)^2$

2. $-4(3)^2$

3. $2(-1)^2$

4. $3(-1)^3$

5. $5(-2)^2$

6. $-\frac{1}{2}(-4)^2$

7. $-2(-3)^2$

8. $-3(-3)^3$

(Answers are on page 156.)

PAINLESS TIP

The order in which you do mathematical operations involving exponents makes a difference in the answer.

For example, $3(2)^2$ is not the same as $(3 \cdot 2)^2$.

The first, $3(2)^2$, is equal to $3(4)$, which is 12.
The second, $(3 \cdot 2)^2$, is equal to 6^2, which is 36.

Adding and Subtracting Exponential Expressions

You can add and subtract exponential expressions if they have the same base and the same exponent. Just add or subtract the coefficients.

Simplify $2(a^3) + 5(a^3)$.
Check to make sure the expressions have the same base and the same exponent.

> The letter a is the base for both.
> The number 3 is the exponent for both.

Next add the coefficients: $2 + 5 = 7$
$2(a^3) + 5(a^3) = 7(a^3)$

Simplify $3(5^2) + 2(5^2)$.
First check to make sure the expressions have the same base and the same exponent.

> The number 5 is the base for both.
> The number 2 is the exponent for both.

Next add the coefficients: $3 + 2 = 5$.
$3(5^2) + 2(5^2) = 5(5^2)$

Simplify $4(y^2) - 2(y^2)$.

Check to make sure the equations have the same base and the same exponent.

> The variable y is the base for both.
> The number 2 is the exponent for both.

Next subtract the coefficients: $4 - 2 = 2$.
$4(y^2) - 2(y^2) = 2y^2$

BRAIN TICKLERS   Set # 36

Simplify each of the following problems by adding or subtracting.

1. $3(3)^2 + 5(3)^2$

2. $4(16)^3 - 2(16)^3$

3. $3x^2 - 5x^2$

4. $2x^0 + 5x^0$

5. $5x^4 - 5x^4$

(Answers are on page 156.)

Multiplying Exponential Expressions

You can multiply two exponential expressions if they have the same base. Just add the exponents.

Simplify $3^3 \cdot 3^2$.

3^3 and 3^2 both have the same base.

To simplify this expression, just add the exponents.

$3^3 \cdot 3^2 = 3^{3+2} = 3^5 = 243$

Simplify $(4)^5(4)^{-3}$.
$(4)^5$ and $(4)^{-3}$ both have the same base.
To simplify this expression, just add the exponents.
$(4)^5(4)^{-3} = 4^{5-3} = 4^2 = 16$

Simplify $(5)^{-10}(5)^{10}$.
$(5)^{-10}$ and $(5)^{10}$ both have the same base.
To simplify this expression, just add the exponents.
$(5)^{-10}(5)^{10} = 5^{-10+10} = 5^0 = 1$

Simplify a^3a^4.
a^3 and a^4 both have the same base.
To simplify this expression, just add the exponents.
$a^3a^4 = a^{3+4} = a^7$

You can multiply several terms. Just add the exponents of all the
terms that have the same base.

Simplify $6^3 \cdot 6^5 \cdot 6^{-2} \cdot 6^4$.
All these terms have the same base.
To simplify this expression, just add all the exponents.
$6^3 \cdot 6^5 \cdot 6^{-2} \cdot 6^4 = 6^{(3+5-2+4)} = 6^{10}$

You can even multiply exponential expressions with coefficients, as
long as they have the same base. Just follow these three *painless* steps.

Step 1: Multiply the coefficients.

Step 2: Add the exponents.

Step 3: Combine the terms. Put the new coefficient first, the base
second, and the new exponent third.

Simplify $3x^24x^5$.
First, multiply the coefficients: $3(4) = 12$.
Second, add the exponents: $2 + 5 = 7$.
Now combine the terms. Put the new coefficient first, the base
second, and the new exponent third.
$3x^24x^5 = 12x^7$

Simplify $-6x(3x^3)$.

First, multiply the coefficients: $(-6)(3) = -18$.

Second, add the exponents: $1 + 3 = 4$.

Now combine the terms. Put the new coefficient first, the base second, and the new exponent third.

$$-6x(3x^3) = -18x^4$$

CAUTION—Major Mistake Territory!

You cannot multiply exponential expressions with different bases.

$3^2 \cdot 2^3$ cannot be simplified, because 3^2 and 2^3 do not have the same base.

a^2b^5 cannot be simplified, because a^2 and b^5 do not have the same base.

BRAIN TICKLERS   Set # 37

Simplify the following exponential expressions.

1. $2^3 2^3$

2. $2^5 2^2$

3. $2^{10} 2^{-2}$

4. $2^{-1} \cdot 2^3 \cdot 2^1$

5. $x^3 x^{-2}$

6. $x^4 \cdot x^{-4}$

7. $6x^4(x^{-2})$

8. $-7x^2(5x^3)$

9. $(-6x^3)(-2x^{-3})$

(Answers are on page 157.)

Dividing Exponential Expressions

You can divide exponential expressions if they have the same base. Just subtract the exponents.

Simplify $3^3 \div 3^2$.
3^3 and 3^2 have the same base.
To simplify, subtract the exponents.
$3^3 \div 3^2 = 3^{3-2} = 3^1 = 3$

Simplify $\dfrac{5^2}{5^3}$.
5^2 and 5^3 have the same base.
To simplify, subtract the exponents.
$\dfrac{5^2}{5^3} = 5^{2-3} = 5^{-1} = \dfrac{1}{5^1} = \dfrac{1}{5}$

Simplify $\dfrac{x^4}{x^{-4}}$.
x^4 and x^{-4} have the same base.
To simplify, subtract the exponents.
$\dfrac{x^4}{x^{-4}} = x^{4-(-4)} = x^8$

Simplify $\dfrac{a^{-3}}{a^2}$.
a^{-3} and a^2 have the same base.
To simplify, subtract the exponents.
$\dfrac{a^{-3}}{a^2} = a^{-3-2} = a^{-5} = \dfrac{1}{a^5}$

You can even divide exponential expressions with coefficients, as long as the expressions have the same base. Just follow these *painless* steps: divide the coefficients, subtract the exponents, and combine the terms. Put the new coefficient first, the base second, and the new exponent third.

Simplify $6x^4 \div 3x^2$.
First divide the coefficients: $6 \div 3 = 2$.
Next subtract the exponents: $4 - 2 = 2$.
Combine the terms. Put the new coefficient first, the base second, and the new exponent third.
$6x^4 \div 3x^2 = 2x^2$

Simplify $\dfrac{4x^3}{16x^{-2}}$.

First divide the coefficients: $\dfrac{4}{16} = \dfrac{1}{4}$.

Next subtract the exponents: $3 - (-2) = 5$.

Combine the terms. Put the new coefficient first, the base second, and the new exponent third.

$$\dfrac{4x^3}{16x^{-2}} = \dfrac{1}{4}x^5$$

⚡ CAUTION—Major Mistake Territory!

You cannot divide exponential expressions with different bases.

$5^2 \div 8^3$ cannot be simplified, because 5^2 and 8^3 do not have the same base.

$\dfrac{a^4}{b^5}$ cannot be simplified, because a^4 and b^5 do not have the same base.

BRAIN TICKLERS   Set # 38

Simplify the following expressions.

1. $\dfrac{2^3}{2^1}$

2. $\dfrac{2^4}{2^{-2}}$

3. $\dfrac{2x^5}{x^5}$

4. $\dfrac{2a^{-2}}{4a^2}$

5. $\dfrac{3x^4}{2x^{-7}}$

(Answers are on page 157.)

Raising to a Power

When you raise an exponential expression to a power, multiply the exponents. Read each of the following examples carefully. Each example illustrates something different about simplifying exponents.

Simplify $(3^4)^5$.
When you raise an exponential expression to a power, just multiply the exponents.
$(3^4)^5 = 3^{(4)(5)} = 3^{20}$

Simplify $(2^2)^{-3}$.
When you raise an exponential expression to a power, multiply the exponents. Remember: a positive number times a negative number is a negative number.
$(2^2)^{-3} = 2^{(2)(-3)} = 2^{-6} = \dfrac{1}{2^6}$

Simplify $(7^{-3})^{-6}$.
Multiply the exponents. Remember: a negative number times a negative number is a positive number.
$(7^{-3})^{-6} = 7^{(-3)(-6)} = 7^{18}$

Simplify $(8^4)^0$.
Multiply the exponents. Remember: any number times zero is zero, and any number to the zero power is 1.
$(8^4)^0 = 8^{(4)(0)} = 8^0 = 1$

BRAIN TICKLERS   Set # 39

Simplify each of the following exponential expressions.

1. $(5^2)^5$

2. $(5^3)^{-1}$

3. $(5^{-2})^{-2}$

4. $(5^4)^0$

5. $(5^2)^3$

(Answers are on page 157.)

Negative Exponents

Sometimes exponential expressions have negative exponents; examples are 5^{-2}, x^{-3}, $2x^{-1}$, and 6^{-2}. In order to solve problems with negative exponents, you have to find the reciprocal of a number.

When you take the reciprocal of a number, you make the numerator the denominator and the denominator the numerator. Basically, you turn a fraction upside down.

Watch—it's easy.

The reciprocal of $\frac{1}{2}$ is $\frac{2}{1}$.

The reciprocal of $\frac{5}{3}$ is $\frac{3}{5}$.

The reciprocal of $-\frac{3}{4}$ is $-\frac{4}{3}$.

What happens when you want to take the reciprocal of a whole number? First change the whole number to an improper fraction, and then take the reciprocal of that fraction.

What is the reciprocal of 5? Because 5 is $\frac{5}{1}$, the reciprocal of 5 is $\frac{1}{5}$.

What is the reciprocal of -3? Because 3 is $-\frac{3}{1}$, the reciprocal of -3 is $-\frac{1}{3}$.

What happens when you want to take the reciprocal of a mixed number? First change the mixed number to an improper fraction, and take the reciprocal of that fraction. It's *painless*.

What is the reciprocal of $5\frac{1}{2}$?

Change $5\frac{1}{2}$ to $\frac{11}{2}$.

The reciprocal of $\frac{11}{2}$ is $\frac{2}{11}$.

The reciprocal of $5\frac{1}{2}$ is $\frac{2}{11}$.

What is the reciprocal of $-6\frac{2}{3}$?

Change $-6\frac{2}{3}$ to $-\frac{20}{3}$.

The reciprocal of $-\frac{20}{3}$ is $-\frac{3}{20}$.

The reciprocal of $-6\frac{2}{3}$ is $-\frac{3}{20}$.

BRAIN TICKLERS   Set # 40

Find the reciprocal of each of the following numbers.

1. 3

2. −8

3. $\frac{4}{3}$

4. $-\frac{2}{3}$

5. $8\frac{1}{2}$

6. $2x$

7. $x-1$

8. $-\frac{x}{3}$

(Answers are on page 157.)

Changing a negative exponent to a positive exponent is a two-step process.

Step 1: Take the reciprocal of the number that is raised to a power.

Step 2: Change the exponent from a negative one to a positive one.

Simplify 5^{-3}.

Step 1: The reciprocal of 5 is $\frac{1}{5}$.

Step 2: Change the exponent from negative three to positive three.

$$5^{-3} = \frac{1}{5^3} = \frac{1}{125}$$

Simplify x^{-3}.

Step 1: The reciprocal of x is $\frac{1}{x}$.

Step 2: Change the exponent from negative three to positive three.

$$x^{-3} = \frac{1}{x^3}$$

Simplify $(x - 2)^{-4}$.

Step 1: The reciprocal of $(x - 2)$ is $\frac{1}{(x-2)}$.

Step 2: Change the exponent from negative four to positive four.

$$(x - 2)^{-4} = \frac{1}{(x-2)^4}$$

Simplify $\left(\frac{3}{5}\right)^{-2}$.

Step 1: The reciprocal of $\frac{3}{5}$ is $\frac{5}{3}$.

Step 2: Change the exponent from negative two to positive two.

$$\left(\frac{3}{5}\right)^{-2} = \left(\frac{5}{3}\right)^2 = \frac{25}{9}$$

BRAIN TICKLERS   Set # 41

Change the following negative exponents to positive exponents.

1. 4^{-3}

2. 3^{-4}

3. 2^{-5}

4. $\left(\frac{2}{5}\right)^{-2}$

5. $\left(\frac{1}{x}\right)^{-3}$

6. $\left(6\frac{1}{2}\right)^{-1}$

7. $(x - 4)^{-2}$

(Answers are on page 158.)

 **SUPER BRAIN TICKLERS**

Compute the value of each of the following exponential expressions.

1. 3^3

2. $\left(\dfrac{1}{3}\right)^2$

3. 3^{-1}

4. $\left(\dfrac{1}{3}\right)^{-2}$

5. $(-3 \cdot 3)^1$

6. $3^2 \cdot 3^{-1}$

7. $\dfrac{3^4}{3^1}$

8. $(3^2)^2$

9. $3^2 + 3^1$

10. $(4 - 1)^{-3}$

(Answers are on page 158.)

Word Problems

Watch how to solve these word problems using exponents.

Problem 1: A number squared plus six squared equals one hundred. What is the number?

Change this problem from Plain English into Math Talk.

"A number squared" becomes "x^2."
"Plus" becomes "$+$."
"Six squared" becomes "6^2."
"Equals" becomes "$=$."
"One hundred" becomes "100."
The result is the equation $x^2 + 6^2 = 100$.

Solve this equation.
$x^2 + 36 = 100$.
Next, subtract 36 from both sides of the equation.
$x^2 + 36 - 36 = 100 - 36$

Simplify $x^2 = 64$.
Take the square root of each side of this equation.
$x = 8$ and $x = -8$

Check your answer. Substitute eight for x.

$8^2 + 6^2 = 100$

$64 + 36 = 100$

The solutions are correct.

Problem 2: Four times a number squared minus that number squared is seventy-five. What is the number?

Change this problem from Plain English into Math Talk.

"Four times a number squared" becomes "$4x^2$."
"Minus" becomes "$-$."
"That number squared" becomes "x^2."
"Is" becomes "$=$."
"Seventy-five" becomes "75."
The result is the equation $4x^2 - x^2 = 75$.

Solve this equation. First, simplify the equation.
$4x^2 - x^2 = 75$ is $3x^2 = 75$

Divide both sides of the equation by three.

$$\frac{3x^2}{3} = \frac{75}{3}$$

Simplify $x^2 = 25$.
Take the square root of each side of the equation.

$x = 5$ and $x = 5$
Check the answers.

$4(5^2) - (5^2) = 75$
$4(25) - 25 = 75$
$100 - 25 = 75$
$75 = 75$

The solution is correct.

BRAIN TICKLERS—THE ANSWERS
Set # 34, page 141

1. $5^2 = 25$
2. $2^6 = 64$
3. $10^2 = 100$
4. $4^3 = 64$
5. $5^0 = 1$

6. $(-3)^2 = 9$
7. $(-5)^3 = -125$
8. $(-4)^2 = 16$
9. $(-1)^5 = -1$
10. $(-1)^{12} = 1$

Set # 35, page 143
1. $3(5)^2 = 3(25) = 75$
2. $-4(3)^2 = -4(9) = -36$
3. $2(-1)^2 = 2(1) = 2$
4. $3(-1)^3 = 3(-1) = -3$
5. $5(-2)^2 = 5(4) = 20$
6. $-\frac{1}{2}(-4)^2 = -\frac{1}{2}(16) = -8$
7. $-2(-3)^2 = -2(9) = -18$
8. $-3(-3)^3 = -3(-27) = 81$

Set # 36, page 145
1. $3(3)^2 + 5(3)^2 = 8(3)^2 = 72$
2. $4(16)^3 - 2(16)^3 = 2(16)^3 = 8{,}192$
3. $3x^2 - 5x^2 = -2x^2$
4. $2x^0 + 5x^0 = 7x^0 = 7$
5. $5x^4 - 5x^4 = 0x^4 = 0$

Set # 37, page 147

1. $2^3 2^3 = 2^6 = 64$

2. $2^5 2^2 = 2^7 = 128$

3. $2^{10} 2^{-2} = 2^8 = 256$

4. $2^{-1} \cdot 2^3 \cdot 2^1 = 2^3 = 8$

5. $x^3 x^{-2} = x^1 = x$

6. $x^4 \cdot x^{-4} = x^0 = 1$

7. $6x^4(x^{-2}) = 6x^2$

8. $-7x^2(5x^3) = -35x^5$

9. $(-6x^3)(-2x^{-3}) = 12x^0 = 12(1) = 12$

Set # 38, page 149

1. $\dfrac{2^3}{2^1} = 2^2 = 4$

2. $\dfrac{2^4}{2^{-2}} = 2^6 = 64$

3. $\dfrac{2x^5}{x^5} = 2x^0 = 2$

4. $\dfrac{2a^{-2}}{4a^2} = \dfrac{1}{2}a^{-4}$

5. $\dfrac{3x^4}{2x^{-7}} = \dfrac{3}{2}x^{11}$

Set # 39, page 150

1. $(5^2)^5 = 5^{10}$

2. $(5^3)^{-1} = 5^{-3}$

3. $(5^{-2})^{-2} = 5^4$

4. $(5^4)^0 = 5^0$

5. $(5^2)^3 = 5^6$

Set # 40, page 152

1. The reciprocal of 3 is $\dfrac{1}{3}$.

2. The reciprocal of -8 is $-\dfrac{1}{8}$.

3. The reciprocal of $\dfrac{4}{3}$ is $\dfrac{3}{4}$.

4. The reciprocal of $-\dfrac{2}{3}$ is $-\dfrac{3}{2}$.

5. The reciprocal of $8\frac{1}{2}$ is $\frac{2}{17}$.

6. The reciprocal of $2x$ is $\frac{1}{2x}$.

7. The reciprocal of $x - 1$ is $\frac{1}{x-1}$.

8. The reciprocal of $-\frac{x}{3}$ is $-\frac{3}{x}$.

Set # 41, page 153

1. $4^{-3} = \left(\frac{1}{4}\right)^3$

2. $3^{-4} = \left(\frac{1}{3}\right)^4$

3. $2^{-5} = \left(\frac{1}{2}\right)^5$

4. $\left(\frac{2}{5}\right)^{-2} = \left(\frac{5}{2}\right)^2$

5. $\left(\frac{1}{x}\right)^{-3} = x^3$

6. $\left(6\frac{1}{2}\right)^{-1} = \frac{2}{13}$

7. $(x - 4)^{-2} = \left(\frac{1}{x-4}\right)^2$

Super Brain Ticklers, page 154

1. $3^3 = 27$

2. $\left(\frac{1}{3}\right)^2 = \frac{1}{9}$

3. $3^{-1} = \frac{1}{3}$

4. $\left(\frac{1}{3}\right)^{-2} = 9$

5. $(-3 \cdot 3)^1 = -9$

6. $3^2 \cdot 3^{-1} = 3$

7. $\frac{3^4}{3^1} = 27$

8. $(3^2)^2 = 81$

9. $3^2 + 3^1 = 12$

10. $(4 - 1)^{-3} = \frac{1}{27}$

Chapter 8

Roots and Radicals

Square Roots

When you square a number, you multiply the number by itself. When you take the square root of a number, you try to figure out, "What number when multiplied by itself will give me this number?"

For example, to figure out, "What is the square root of twenty-five?," rephrase the question as "What number when multiplied by itself equals twenty-five?" The answer is five. Five times five equals twenty-five. Five is the square root of twenty-five.

To figure out, "What is the square root of one hundred?," think of the question as "What number when multiplied by itself equals one hundred?" The answer is ten. Ten times ten equals one hundred. Ten is the square root of one hundred.

You write a square root by putting a number under a *radical sign*. A radical sign looks like this: $\sqrt{}$. When you see a radical sign, take the square root of the number under it. The number under the radical sign is called a *radicand*. Look at $\sqrt{5}$. Five is under the radical sign. Five is the radicand.

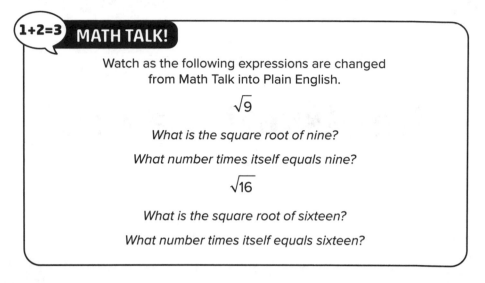

1+2=3 **MATH TALK!**

Watch as the following expressions are changed
from Math Talk into Plain English.

$$\sqrt{9}$$

What is the square root of nine?

What number times itself equals nine?

$$\sqrt{16}$$

What is the square root of sixteen?

What number times itself equals sixteen?

After you figure out the square root of a number, you can check your
answer by multiplying the number by itself to see if the answer is the
radicand. For example, what is the square root of nine? The square
root of nine is three. To check, multiply three times three. Three
times three is nine. The answer is correct.

Certain numbers are *perfect squares.* The square root of a perfect
square is a whole number. For example, 16 and 25 are both perfect
squares.

Every number has both a positive and a negative square root. When
the square root is written without any sign in front of it, the answer
is the positive square root. When the square root is written with a
negative sign in front of it, the answer is the negative square root.

16 is a perfect square.
16 has both a positive and a negative square root.
$\sqrt{16}$ is 4, since $4(4) = 16$.
    The negative square root of 16 is -4, since $(-4)(-4) = 16$.

25 is a perfect square.
25 has both a positive and a negative square root.
$\sqrt{25}$ is 5, since $(5)(5) = 25$.

    The negative square root of 25 is -5, since $(-5)(-5) = 25$.

PERFECT SQUARE ROOTS

Memorize all of these.

What is $\sqrt{0}$?   It is 0.   Check it. Square 0. $0^2 = 0 \cdot 0 = 0$

What is $\sqrt{1}$?   It is 1.   Check it. Square 1. $1^2 = 1 \cdot 1 = 1$

What is $\sqrt{4}$?   It is 2.   Check it. Square 2. $2^2 = 2 \cdot 2 = 4$

What is $\sqrt{9}$?   It is 3.   Check it. Square 3. $3^2 = 3 \cdot 3 = 9$

What is $\sqrt{16}$?   It is 4.   Check it. Square 4. $4^2 = 4 \cdot 4 = 16$

What is $\sqrt{25}$?   It is 5.   Check it. Square 5. $5^2 = 5 \cdot 5 = 25$

What is $\sqrt{36}$?   It is 6.   Check it. Square 6. $6^2 = 6 \cdot 6 = 36$

What is $\sqrt{49}$?   It is 7.   Check it. Square 7. $7^2 = 7 \cdot 7 = 49$

What is $\sqrt{64}$?   It is 8.   Check it. Square 8. $8^2 = 8 \cdot 8 = 64$

What is $\sqrt{81}$?   It is 9.   Check it. Square 9. $9^2 = 9 \cdot 9 = 81$

What is $\sqrt{100}$? It is 10.   Check it. Square 10. $10^2 = 10 \cdot 10 = 100$

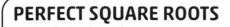

BRAIN TICKLERS   Set # 42

Quickly solve these square root problems.

1. $\sqrt{9}$

2. $\sqrt{25}$

3. $\sqrt{36}$

4. $\sqrt{81}$

5. $\sqrt{49}$

6. $\sqrt{100}$

7. $\sqrt{16}$

8. $\sqrt{4}$

9. $\sqrt{64}$

10. $\sqrt{1}$

(Answers are on page 190.)

The radicands in the preceding Brain Ticklers are all perfect squares. Some numbers are not perfect squares. If a number is not a perfect square, no *whole* number when multiplied by itself will equal that number.

5 is not a perfect square; $\sqrt{5}$ is not a whole number.

3 is not a perfect square; $\sqrt{3}$ is not a whole number.

Cube Roots and Higher

When a number is multiplied by itself three times, it is *cubed*. Two cubed is two times two times two. Two cubed is eight. Taking the cube root of a number is the opposite of cubing a number. "What is the cube root of eight?" really asks "What number when multiplied by itself three times equals eight?" The answer is two. Two times two times two is equal to eight. Two is the cube root of eight.

To write a *cube root*, draw a radical sign and put the number three inside the crook of the radical sign. The expression $\sqrt[3]{8}$ is read as "What is the cube root of eight?" The number three is called the *index*. It tells you how many times the number must be multiplied by itself.

Notice the difference between these four sentences and their mathematical equivalents.

The cube root of eight is two. $\sqrt[3]{8} = 2$

Two is the cube root of eight. $2 = \sqrt[3]{8}$

Two cubed is eight. $2^3 = 8$

Eight is two cubed. $8 = 2^3$

1+2=3 MATH TALK!

In the expression $\sqrt[3]{8}$,

$\sqrt{}$ is the radical sign;

3 is the index;

8 is the radicand.

It is also possible to take the fourth root of a number. To ask, "What is the fourth root of a number?" in mathematical terms, write four as the index in the radical sign: $\sqrt[4]{}$. The radical expression now poses the question "What number when multiplied by itself four times equals the number under the radical sign?" The radical expression $\sqrt[4]{16}$ is read as "What is the fourth root of sixteen?" The problem can be rephrased as "What number when multiplied by itself four times is equal to sixteen?" The answer is two. Two times two is four, times two is eight, times two is sixteen.

Two to the fourth power is sixteen. $2^4 = 16$

The fourth root of sixteen is two. $\sqrt[4]{16} = 2$

The index of a radical can also be a natural number greater than four. If the index is ten, it asks, "What number multiplied by itself ten times is equal to the number under the radical sign?" If the index is 100, it asks, "What number when multiplied by itself 100 times is equal to the number under the radical sign?"

1+2=3 MATH TALK!

The following radical expressions are changed from Math Talk into Plain English.

$$\sqrt[3]{27} = 3$$

The cube root of twenty-seven is three.

$$3 = \sqrt[3]{27}$$

Three is the cube root of twenty-seven.

$$\sqrt[4]{16} = 2$$

The fourth root of sixteen is two.

$$2 = \sqrt[4]{16}$$

Two is the fourth root of sixteen.

$$\sqrt[6]{64} = 2$$

The sixth root of sixty-four is two.

$$2 = \sqrt[6]{64}$$

Two is the sixth root of sixty-four.

BRAIN TICKLERS   Set # 43

Solve the following radical expressions by finding the positive roots.

1. $\sqrt[3]{27}$

2. $\sqrt[3]{64}$

3. $\sqrt[5]{1}$

4. $\sqrt[4]{16}$

5. $\sqrt[3]{125}$

6. $\sqrt[10]{0}$

(Answers are on page 190.)

Sometimes the index of a radical is a variable.

For example, $\sqrt[x]{9}$, $\sqrt[y]{16}$, $\sqrt[x]{25}$, $\sqrt[y]{32}$.

1+2=3 MATH TALK!

Watch as these radical expressions are changed from Math Talk into Plain English.

$$\sqrt[x]{9}$$

What is the xth root of nine?
What number multiplied by itself
x times equals nine?

$$\sqrt[y]{16}$$

What is the yth root of sixteen?
What number multiplied by itself y times
equals sixteen?

It is impossible to compute the value of a radical expression when the index is a variable. For example, it is impossible to compute the value of $\sqrt[y]{16}$ unless you know the value of y. If $y = 2$, the value of $\sqrt[y]{16}$ is 4. But if $y = 4$, the value of $\sqrt[y]{16}$ is 2.

It is possible to solve for the index if you know the value of the radical expression. To solve $\sqrt[x]{9} = 3$ for x, rewrite this radical expression as an exponential expression.

Rewrite $\sqrt[x]{9} = 3$ as $3^x = 9$.

Solve for x by substituting different numbers for x.
If $x = 1$, then $3^x = 3^1 = 3$.
If $x = 2$, $3^x = 3^2 = 9$.
$x = 2$

Solve for x: $\sqrt[x]{125} = 5$

Rewrite $\sqrt[x]{125}$ as the exponential expression $5^x = 125$.
Substitute different natural numbers for x. Start with one.
If $x = 1$, $5^x = 5^1 = 5$; x is not equal to one.
If $x = 2$, $5^x = 5^2 = 25$; x is not equal to two.
If $x = 3$, $5^x = 5^3 = 125$.
$x = 3$

BRAIN TICKLERS   Set # 44

Solve for x.

1. $\sqrt[x]{16} = 4$

2. $\sqrt[x]{16} = 2$

3. $\sqrt[x]{25} = 5$

4. $\sqrt[x]{125} = 5$

5. $\sqrt[x]{8} = 2$

(Answers are on page 190.)

Negative Radicands

What happens when the number under the radical sign is a negative number?

For example:

$\sqrt{-4}$ What is the square root of negative four?

$\sqrt[3]{-8}$ What is the cube root of negative eight?

$\sqrt[4]{-16}$ What is the fourth root of negative sixteen?

$\sqrt[5]{-32}$ What is the fifth root of negative thirty-two?

The answer depends on whether the index is even or is odd.

Case 1: The index is even.

When the index is even, you cannot compute the value of a negative radicand. Watch.

What is $\sqrt{-4}$?

Try to find the square root of negative four.
What number when multiplied by itself equals negative four?

$$(+2)(+2) = +4, \text{ not } -4.$$
$$(-2)(-2) = +4, \text{ not } -4.$$

There is no real number that when multiplied by itself equals negative four.

What is $\sqrt[4]{-16}$?

$$(+2)(+2)(+2)(+2) = +16, \text{ not } -16.$$
$$(-2)(-2)(-2)(-2) = +16, \text{ not } -16.$$

There is no real number that when multiplied by itself four times equals negative sixteen.
The even root of a negative number is always undefined in the real number system.

$$\sqrt[4]{-81} \text{ is undefined.}$$
$$\sqrt{-25} \text{ is undefined.}$$
$$\sqrt[10]{-20} \text{ is undefined.}$$

$\sqrt[8]{-100}$ is undefined.

$\sqrt[100]{-1}$ is undefined.

CAUTION—Major Mistake Territory!

$\sqrt{-9}$ is not negative three.

$\sqrt[4]{-16}$ is not negative two.

When the index of a radical is even, you cannot take the root of a negative number.

Case 2: The index is odd.

What is $\sqrt[3]{-8}$?

The expression $\sqrt[3]{-8}$ asks, "What number when multiplied by itself three times is equal to negative eight?" The answer is negative two.

$$(-2)(-2)(-2) = -8$$

Negative two times negative two times negative two is equal to negative eight. Why?

Multiply the first two negative twos.

$$(-2)(-2) = 4$$

Negative two times negative two equals positive four. Now multiply four by negative two.

$$(4)(-2) = (-8)$$

Four times negative two is negative eight.

When you cube a negative number, the answer is negative.

$$\sqrt[3]{-8} = -2$$

If the index of a radical is odd, the problem has a solution. If the number under the radical sign is positive, the solution is positive. If the number under the radical sign is negative, the solution is negative.

Watch carefully what happens when -1 is under the radical sign.

$$\sqrt{-1} \text{ is undefined.}$$
$$\sqrt[3]{-1} = -1$$
$$\sqrt[25]{-1} = -1$$
$$\sqrt[100]{-1} \text{ is undefined.}$$

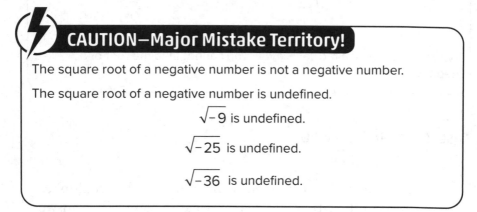

CAUTION—Major Mistake Territory!

The square root of a negative number is not a negative number.

The square root of a negative number is undefined.

$$\sqrt{-9} \text{ is undefined.}$$

$$\sqrt{-25} \text{ is undefined.}$$

$$\sqrt{-36} \text{ is undefined.}$$

BRAIN TICKLERS   Set # 45

Solve the following radicals. Be careful. Some of them are undefined.

1. $\sqrt{-64}$

2. $\sqrt[4]{-16}$

3. $\sqrt[3]{-27}$

4. $\sqrt[5]{-32}$

5. $\sqrt{-49}$

6. $\sqrt[9]{-1}$

(Answers are on page 190.)

Radical Expressions
Simplifying radical expressions

There are rules for simplifying radical expressions.

Rule 1: If two numbers are multiplied under a radical sign, you can rewrite them under two different radical signs. The two expressions are then multiplied.

$$\sqrt{(9)(16)} = \left(\sqrt{9}\right)\left(\sqrt{16}\right)$$

$$\sqrt{(25)(4)} = \left(\sqrt{25}\right)\left(\sqrt{4}\right)$$

$$\sqrt{(36)(100)} = \left(\sqrt{36}\right)\left(\sqrt{100}\right)$$

Watch how separating a radical expression into two separate expressions makes simplifying easy.

Simplify $\sqrt{(9)(16)}$.

Rewrite $\sqrt{(9)(16)}$ as two separate expressions.

$$\sqrt{(9)(16)} = \left(\sqrt{9}\right)\left(\sqrt{16}\right)$$

Solve each of the two radical expressions.

$$\sqrt{9} = 3 \text{ and } \sqrt{16} = 4$$

Multiply the solutions.

$$(3)(4) = 12$$

The result is the answer.

$$\sqrt{(9)(16)} = 12$$

Simplify $\sqrt{(4)(16)(25)}$.

Rewrite $\sqrt{(4)(16)(25)}$ as three separate expressions.

$$\sqrt{(4)(16)(25)} = \left(\sqrt{4}\right)\left(\sqrt{16}\right)\left(\sqrt{25}\right)$$

Solve each of the three radical expressions.

$$\sqrt{4} = 2 \text{ and } \sqrt{16} = 4 \text{ and } \sqrt{25} = 5$$

Multiply the solutions.

$$(2)(4)(5) = 40$$

The result is the answer.

$$\sqrt{(4)(16)(25)} = 40$$

Sometimes one of the numbers under the radical sign is not a perfect square. In that case, take the square root of the number that is a perfect square and multiply the result by the square root of the number that is not a perfect square. Does this sound complicated? It's not. It's *painless.*

Simplify $\sqrt{(25)(6)}$.

Rewrite $\sqrt{(25)(6)}$ as two separate expressions.

$$\sqrt{(25)(6)} = \left(\sqrt{25}\right)\left(\sqrt{6}\right)$$

Take the square root of the expression that is a perfect square.

$$\sqrt{25} = 5; \qquad \sqrt{6} \text{ is not a perfect square.}$$

Multiply the solutions.

$$5\sqrt{6}$$

The result is the answer.

$$\sqrt{(25)(6)} = 5\sqrt{6}$$

It isn't necessary to find the actual value of $\sqrt{6}$.

Simplify $\sqrt{49\,x}$.

Rewrite.

$$\sqrt{49} \cdot \sqrt{x}$$

Take the square root of the expression that is a perfect square.

$$\sqrt{49} = 7; \qquad \sqrt{x} \text{ is not a perfect square.}$$

Multiply the two solutions.

$$7\sqrt{x}$$

The result is the answer.

$$\sqrt{49x} = 7\sqrt{x}$$

BRAIN TICKLERS   Set # 46

Solve each of the following square root problems by rewriting the radical expression as two separate expressions.

1. $\sqrt{(16)(9)}$

2. $\sqrt{(64)(100)}$

3. $\sqrt{25y^2}$ $(y > 0)$

4. $\sqrt{(4)(11)}$

5. $\sqrt{9y}$

(Answers are on page 191.)

Rule 2: If two radical expressions are multiplied, you can rewrite them as products under the same radical sign.

$\left(\sqrt{27}\right)\left(\sqrt{3}\right)$ can be rewritten as $\sqrt{(27)(3)}$.

$\left(\sqrt{8}\right)\left(\sqrt{5}\right)$ can be rewritten as $\sqrt{(8)(5)}$.

Rewriting two radical expressions that are multiplied under the same radical sign can often help simplify them. Watch.

Simplify $\left(\sqrt{27}\right)\left(\sqrt{3}\right)$.

It is impossible to remove the radical sign from either $\sqrt{27}$ or $\sqrt{3}$, so these expressions cannot be simplified further.

Put both of these radical expressions under the same radical sign.

$$\left(\sqrt{27}\right)\left(\sqrt{3}\right) = \sqrt{(27)(3)}$$

Multiply the expression under the radical sign.

$$\sqrt{(27)(3)} = \sqrt{81}$$

81 is a perfect square.

$$\sqrt{81} = 9$$

The result is the answer.

$$\left(\sqrt{27}\right)\left(\sqrt{3}\right) = 9$$

Simplify $\left(\sqrt{5}\right)\left(\sqrt{5}\right)$.

It is impossible to simplify each $\sqrt{5}$.
Put both of these radical expressions under the same radical sign.

$$\left(\sqrt{5}\right)\left(\sqrt{5}\right) = \sqrt{(5)(5)}$$

Multiply the expression under the radical sign.

$$\sqrt{(5)(5)} = \sqrt{25}$$

25 is a perfect square.

$$\sqrt{25} = 5$$

The result is the solution.

$$\left(\sqrt{5}\right)\left(\sqrt{5}\right) = 5$$

BRAIN TICKLERS   Set # 47

In each case, simplify the radical expressions by placing them under the same radical sign.

1. $\left(\sqrt{3}\right)\left(\sqrt{3}\right)$

4. $\left(\sqrt{x}\right)\left(\sqrt{x}\right)$ $(x > 0)$

2. $\left(\sqrt{8}\right)\left(\sqrt{2}\right)$

5. $\left(\sqrt{x^3}\right)\left(\sqrt{x}\right)$ $(x > 0)$

3. $\left(\sqrt{(12)}\right)\left(\sqrt{3}\right)$

(Answers are on page 191.)

Factoring a radical expression

Sometimes the number under the square root sign is not a perfect square, but it can still be simplified.

Rule 3: You can factor the number under a radical and take the square root of one of the factors.

For example, what is $\sqrt{12}$?
What number when multiplied by itself is equal to twelve?
There is no whole number that when multiplied by itself equals twelve. But if you use a calculator, enter the number twelve, and tap the square root symbol, 3.464101615138 will show up on the screen.
$(3.464101615138)(3.464101615138) = 12$

In addition to using a calculator to compute square roots, mathematicians often simplify them. To simplify a square root, look at the factors of the number under the radical sign.

Simplify $\sqrt{12}$.

What are the factors of 12?
$$(3)(4) = 12$$
$$(2)(6) = 12$$

Are any of the factors perfect squares?
Out of all of these numbers only four is a perfect square.
$$(2)(2) = 4$$

Rewrite $\sqrt{12}$ as $\sqrt{(4)(3)}$.

Rewrite $\sqrt{(4)(3)}$ as $\left(\sqrt{4}\right)\left(\sqrt{3}\right)$.

Now $\sqrt{4} = 2$, so rewrite $\left(\sqrt{4}\right)\left(\sqrt{3}\right)$ as $2\sqrt{3}$.
$\sqrt{12}$ is $2\sqrt{3}$.

Simplify $\sqrt{18}$.
Find the factors of 18.
$$(9)(2) = 18$$
$$(3)(6) = 18$$
Are any of the factors perfect squares?

Nine is a perfect square, so $\sqrt{18}$ can be simplified.
Rewrite $\sqrt{18}$ as $\left(\sqrt{9}\right)\left(\sqrt{2}\right)$.
Now $\sqrt{9} = 3$, so rewrite $\left(\sqrt{9}\right)\left(\sqrt{2}\right)$ as $3\sqrt{2}$.
$\sqrt{18}$ is $3\sqrt{2}$.

BRAIN TICKLERS   Set # 48

Simplify the following radical expressions by factoring.

1. $\sqrt{20}$

2. $\sqrt{8}$

3. $\sqrt{27}$

4. $\sqrt{24}$

5. $\sqrt{32}$

6. $\sqrt{125}$

(Answers are on page 191.)

Division of radicals

Rule 4: If two numbers are divided under a radical sign, you can rewrite them under two different radical signs separated by a division sign.

$\sqrt{\dfrac{9}{4}}$ can be rewritten as $\dfrac{\sqrt{9}}{\sqrt{4}}$.

$\sqrt{\dfrac{25}{36}}$ can be rewritten as $\dfrac{\sqrt{25}}{\sqrt{36}}$.

$\sqrt{\dfrac{64}{16}}$ can be rewritten as $\dfrac{\sqrt{64}}{\sqrt{16}}$.

To take the square root of a rational number, rewrite the problem by putting the numerator and denominator under two different radical signs. Next take the square root of each one of the numbers. Watch. It's *painless*.

Simplify $\sqrt{\dfrac{9}{4}}$.

Rewrite the problem as $\dfrac{\sqrt{9}}{\sqrt{4}}$.

$$\sqrt{\dfrac{9}{4}} = \dfrac{3}{2}$$

Simplify $\sqrt{\dfrac{x^2}{16}}$.

Rewrite the problem as $\dfrac{\sqrt{x^2}}{\sqrt{16}}$.

In this example, $x > 0$.

$$\sqrt{\dfrac{x^2}{16}} = \dfrac{x}{4}$$

Simplify $\sqrt{\dfrac{3}{25}}$.

Rewrite the problem as $\dfrac{\sqrt{3}}{\sqrt{25}}$.

$\sqrt{3}$ cannot be simplified. $\sqrt{25} = 5$

$$\sqrt{\dfrac{3}{25}} = \dfrac{\sqrt{3}}{5}$$

Simplify $\sqrt{\dfrac{49}{5}}$.

Rewrite the problem as $\dfrac{\sqrt{49}}{\sqrt{5}}$.

$$\sqrt{\dfrac{49}{5}} = \dfrac{7}{\sqrt{5}}$$

However, the answer, $\dfrac{7}{\sqrt{5}}$, has a radical in the denominator.

If an answer has a radical in the denominator, the expression is not considered simplified. In order to simplify this expression, you have to rationalize the denominator. Let's see how.

Rationalizing the denominator

A radical expression is not simplified if there is a radical in the denominator.

$\dfrac{3}{\sqrt{2}}$ is not simplified because $\sqrt{2}$ is in the denominator.

$\dfrac{\sqrt{5}}{\sqrt{3}}$ is not simplified because $\sqrt{3}$ is in the denominator.

$\dfrac{5}{\sqrt{x}}$ is not simplified because $\sqrt{x}$ is in the denominator.

Rule 5: You can multiply the numerator and denominator of a radical expression by the same number without changing the value of the expression.

To eliminate a radical expression in the denominator, follow these *painless* steps.

Step 1: Identify the radical expression in the denominator.

Step 2: Construct a fraction in which this radical expression is both the numerator and the denominator.

Step 3: Multiply the original expression by the fraction.

Step 4: Put the two square roots in the denominator under the same radical.

Step 5: Take the square root of the number in the denominator. The result is the answer.

Does this sound complicated? Watch. It's *painless*.

Rationalize the denominator in the expression $\dfrac{3}{\sqrt{2}}$.

Step 1: Identify the radical expression in the denominator.

The radical expression in the denominator is $\sqrt{2}$.

Step 2: Construct a fraction in which this radical expression is both the numerator and the denominator. The value of the fraction is 1.

$$\frac{\sqrt{2}}{\sqrt{2}} = 1$$

Step 3: Multiply the original expression by the new expression.

$$\left(\frac{3}{\sqrt{2}}\right)\left(\frac{\sqrt{2}}{\sqrt{2}}\right) \frac{3\sqrt{2}}{\sqrt{2}\sqrt{2}}$$

Step 4: Put the two square roots in the denominator under the same radical.

$$\frac{3\sqrt{2}}{\sqrt{2}\sqrt{2}} = \frac{3\sqrt{2}}{\sqrt{2\cdot 2}} = \frac{3\sqrt{2}}{\sqrt{4}}$$

Step 5: Take the square root of the number in the denominator. The denominator is $\sqrt{4}$.

$$\frac{3\sqrt{2}}{\sqrt{4}} = \frac{3\sqrt{2}}{2}$$

The result is the solution.

$$\frac{3}{\sqrt{2}} = \frac{3\sqrt{2}}{2}$$

Rationalize the denominator in the expression $\sqrt{\dfrac{25}{3}}$.

Step 1: Identify the radical expression in the denominator.

$$\sqrt{\frac{25}{3}} = \frac{\sqrt{25}}{\sqrt{3}} = \frac{5}{\sqrt{3}}$$

$\sqrt{3}$ is the radical expression in the denominator.

Step 2: Construct a fraction in which this radical expression is both the numerator and the denominator.

$$\frac{\sqrt{3}}{\sqrt{3}} = 1$$

Step 3: Multiply the original expression by this fraction.

$$\frac{5}{\sqrt{3}} \cdot \frac{\sqrt{3}}{\sqrt{3}} = \frac{5\sqrt{3}}{\sqrt{3}\sqrt{3}}$$

Step 4: Put the two square roots in the denominator under the same radical.

$$\frac{5\sqrt{3}}{\sqrt{3}\sqrt{3}} = \frac{5\sqrt{3}}{\sqrt{3 \cdot 3}} = \frac{5\sqrt{3}}{\sqrt{9}}$$

Step 5: Take the square root of the number in the denominator. The denominator is $\sqrt{9}$. The $\sqrt{9}$ is 3.

$$\frac{5\sqrt{3}}{\sqrt{9}} = \frac{5\sqrt{3}}{3}$$

The result is the solution.

$$\sqrt{\frac{25}{3}} = \frac{5\sqrt{3}}{3}$$

BRAIN TICKLERS   Set # 49

Rationalize the denominator in each of the following expressions.

1. $\dfrac{5}{\sqrt{3}}$

2. $\dfrac{10}{\sqrt{2}}$

3. $\sqrt{\dfrac{16}{3}}$

4. $\sqrt{\dfrac{36}{7}}$

(Answers are on page 191.)

Adding and subtracting radical expressions

Rule 6: You can add and subtract radical expressions if the indexes are the same and the numbers under the radical sign are the same. When you add radical expressions, add the coefficients.

$2\sqrt{3} + 3\sqrt{3}$

First check to make sure the radicals have the same index. No index is written in either of these expressions, so the index in both of them is two.

Next check if both radical expressions have the same quantity under the radical sign. Both expressions have the number 3 under the radical sign. The number 3 is the radicand.

Add $2\sqrt{3}$ and $3\sqrt{3}$ by adding the coefficients.

The coefficient of $2\sqrt{3}$ is 2.

The coefficient of $3\sqrt{3}$ is 3.

Add the coefficients: $2 + 3 = 5$.

Attach the radical expression, $\sqrt{3}$, to the new coefficient.

$2\sqrt{3} + 3\sqrt{3} = 5\sqrt{3}$.

$\sqrt[3]{5} + 5\sqrt[3]{5}$

First check to make sure the radicals have the same index. Three is the index in both expressions.

Next check to see if both radical expressions have the same quantity under the radical sign. Both expressions have the number 5 under the radical sign.

Since both expressions have the same index and the same expression under the radical sign, they can be added.

To add $\sqrt[3]{5}$ and $5\sqrt[3]{5}$, just add the coefficients.

The coefficient of $\sqrt[3]{5}$ is 1.

The coefficient of $5\sqrt[3]{5}$ is 5.

Add the coefficients: $1 + 5 = 6$.

Attach the radical expression, $\sqrt[3]{5}$, to the new coefficient.

$\sqrt[3]{5} + 5\sqrt[3]{5} = 6\sqrt[3]{5}$.

$\sqrt{x} - 4\sqrt{x}$

First check to make sure the radicals have the same index. The index in both of them is two.

Next check if both radical expressions have the same quantity under the radical sign. Both expressions have x under the radical sign, so one can be subtracted from the other.

To subtract $\sqrt{x} - 4\sqrt{x}$, just subtract the coefficients.

The coefficient of $\sqrt{x}$ is 1.

The coefficient of $4\sqrt{x}$ is 4.

Subtract the coefficients: $1 - 4 = -3$.

Attach the radical expression, $\sqrt{x}$, to the new coefficient.

$\sqrt{x} - 4\sqrt{x} = -3\sqrt{x}$.

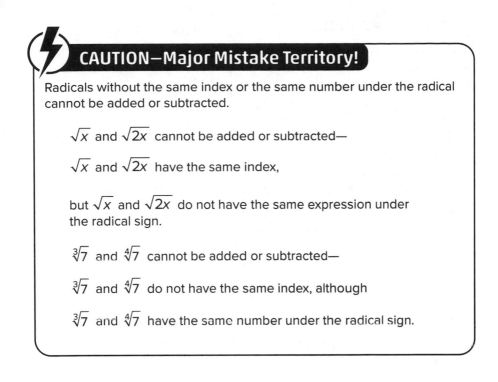

CAUTION—Major Mistake Territory!

Radicals without the same index or the same number under the radical cannot be added or subtracted.

$\sqrt{x}$ and $\sqrt{2x}$ cannot be added or subtracted—

$\sqrt{x}$ and $\sqrt{2x}$ have the same index,

but $\sqrt{x}$ and $\sqrt{2x}$ do not have the same expression under the radical sign.

$\sqrt[3]{7}$ and $\sqrt[4]{7}$ cannot be added or subtracted—

$\sqrt[3]{7}$ and $\sqrt[4]{7}$ do not have the same index, although

$\sqrt[3]{7}$ and $\sqrt[4]{7}$ have the same number under the radical sign.

BRAIN TICKLERS   Set # 50

In each of the following, add or subtract the radical expressions.

1. $7\sqrt{2} + \sqrt{2}$

2. $3\sqrt{3} + \sqrt{3}$

3. $5\sqrt[4]{x} + 2\sqrt[4]{x}$

4. $9\sqrt{5} - \sqrt{5}$

5. $2\sqrt{x} - 2\sqrt{x}$

6. $5\sqrt{2x} - 12\sqrt{2x}$

(Answers are on page 192.)

Fractional exponents

Rule 7: Radical expressions can be written as fractional exponents. The numerator of the exponent is the power of the radicand. The denominator of the exponent is the index. Watch.

$\sqrt[3]{x^2}$ is the same as $x^{\frac{2}{3}}$.

Follow these simple steps to change a radical expression into an exponential one.

Step 1: Copy the number or expression under the radical sign.

Step 2: Make the exponent of the number or expression the numerator of the expression.

Step 3: Make the index the denominator of the expression.

Change $\sqrt[3]{2^5}$ to an exponential expression.

Step 1: Copy the number or expression under the radical sign.

Two is the base of the radicand. Copy the two.

$$2$$

Step 2: Make the exponent of the base the numerator of the expression.

Five is the exponent. Make five the numerator.

$$2^{\frac{5}{?}}$$

Step 3: Make the index the denominator of the expression.

Three is the index. Make three the denominator.

$$2^{\frac{5}{3}}$$

Answer: $\sqrt[3]{2^5} = 2^{\frac{5}{3}}$

Change $\sqrt{5}$ to an exponential expression.

Step 1: Copy the base under the radical sign.
Copy the five.

$$5$$

Step 2: Make the exponent of the radicand the numerator of the expression.
The exponent is one. Make one the numerator.

$$5^{\frac{1}{?}}$$

Step 3: Make the index the denominator of the expression.
Since no index is written, the index is two. Make two the denominator.

$$5^{\frac{1}{2}}$$

Answer: $\sqrt{5} = 5^{\frac{1}{2}}$

Change $\sqrt{(3xy)^3}$ to an exponential expression.

Step 1: Copy the base under the radical sign.
Put the entire expression in parentheses. Put the expression $3xy$ in parentheses.

$$(3xy)$$

Step 2: Make the exponent the numerator of the expression.
Three is the exponent. Make three the numerator.

$$(3xy)^{\frac{3}{?}}$$

Step 3: Make the index the denominator of the expression.
The index is two. Make two the denominator.

$$(3xy)^{\frac{3}{2}}$$

Answer: $\sqrt{(3xy)^3} = (3xy)^{\frac{3}{2}}$

Rule 8: An exponential expression with a fraction as the exponent can be changed into a radical expression. The numerator of the exponent is the power of the radicand. The denominator of the exponent is the index of the radical expression.

To change an exponential expression with a fractional exponent into a radical expression, follow these three *painless* steps.

Step 1: Write the base of the exponential expression under a radical sign.

Step 2: Raise the number under the radical sign to the power of the numerator of the fractional exponent.

Step 3: Make the denominator of the fractional exponent the index of the radical.

It's *painless*. Really!

Change $x^{\frac{1}{3}}$ to a radical expression.

Step 1: Write the base of the exponential expression under a radical sign.
Write x under a radical sign.

$$\sqrt{x}$$

Step 2: Raise the number under the radical sign to the power of the numerator of the fractional exponent.
Raise x to the first power. Because one is the numerator of the exponent, you don't have to write it.

$$\sqrt{x^1} = \sqrt{x}$$

Step 3: Make the denominator of the fractional exponent the index of the radical.

$$\sqrt[3]{x}$$

Answer: $x^{\frac{1}{3}} = \sqrt[3]{x}$

Change $(3xy)^{\frac{3}{2}}$ to a radical expression.

Step 1: Write the base of the exponential expression under a radical sign.

The quantity $3xy$ is raised to the $\frac{3}{2}$ power, so $3xy$ is the base of the expression.

$$\sqrt{3xy}$$

Step 2: Raise the number under the radical sign to the power of the numerator of the fractional exponent.

Raise $3xy$ to the third power. Be careful—$\sqrt{3xy^3}$ is not the same as $\sqrt{(3xy)^3}$.

$$\sqrt{(3xy)^3}$$

Step 3: Make the denominator of the fractional exponent the index of the radical.

Make two the index.

$$\sqrt[2]{(3xy)^3}$$

When two is the index, you don't need to write it.

$$\sqrt[2]{(3xy)^3} = \sqrt{(3xy)^3}$$

Answer: $(3xy)^{\frac{3}{2}} = \sqrt{(3xy)^3}$

REMINDER

Remember to put parentheses around the radicand.

$$3x^2 \neq (3x)^2$$
$$12x^3 \neq (12x)^3$$

BRAIN TICKLERS   Set # 51

Change the radical expressions into exponential expressions.

1. $\sqrt{5}$

2. $\sqrt{x}$

3. $\sqrt[3]{12}$

4. $\sqrt[5]{7^2}$

5. $\sqrt[3]{(2x)^2}$

(Answers are on page 192.)

BRAIN TICKLERS   Set # 52

Change the exponential expressions into radical expressions.

1. $10^{\frac{1}{2}}$

2. $3^{\frac{1}{3}}$

3. $(5xy)^{\frac{1}{2}}$

4. $7^{\frac{2}{3}}$

5. $(9x)^{\frac{2}{3}}$

(Answers are on page 192.)

SUPER BRAIN TICKLERS

Solve.

1. $\sqrt[3]{125}$

2. $\sqrt[3]{-27}$

3. $\sqrt{12}$

4. $\sqrt{-49}$

5. $\sqrt{(8)(2)}$

6. $\left(\sqrt{12}\right)\left(\sqrt{3}\right)$

7. $\sqrt{32}$

8. $\dfrac{8}{\sqrt{5}}$

9. $2\sqrt{3} - 4\sqrt{3}$

10. $\left(5\sqrt{2}\right)\left(2\sqrt{18}\right)$

(Answers are on page 192.)

Word Problems

Watch as the following word problems that use roots and radicals are solved.

Problem 1: The square root of a number plus two times the square root of the same number is twelve. What is the number?

First change this problem from Plain English into Math Talk.
"The square root of a number" becomes "$\sqrt{x}$."
"Plus" becomes "+."
"Two times the square root of the same number" becomes "$2\sqrt{x}$."
"Is" becomes "=."
"Twelve" becomes "12."
Now you can change this problem into an equation.
$$\sqrt{x} + 2\sqrt{x} = 12$$

Now solve this equation.
Add $\sqrt{x} + 2\sqrt{x}$. You can add these two expressions because they have the same radicand and the same index.
$$\sqrt{x} + 2\sqrt{x} = 3\sqrt{x}$$

The new equation is $3\sqrt{x} = 12$.
Divide both sides of the equation by 3.

$$\frac{3\sqrt{x}}{3} = \frac{12}{3}$$

Compute.

$$\sqrt{x} = 4$$

Square both sides of the equation.

$$\left(\sqrt{x}\right)^2 = (4)^2$$

Compute.

$x = 16$
The answer is 16.

Problem 2: Seth lives in Bethesda. Sarah lives nine miles east of Seth. Joanne lives twelve miles north of Sarah. How far does Seth live from Joanne?

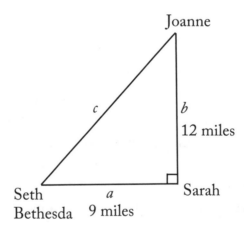

The first step is to change this problem from Plain English into Math Talk. To do this, make a drawing of the problem. The drawing should be a right triangle. The distance from Seth's house to Joanne's house is the hypotenuse of the right triangle.

To find this distance use the Pythagorean Theorem
$a^2 + b^2 = c^2$.

> a is the distance from Seth's house to Sarah's house, which is 9 miles.
>
> b is the distance from Sarah's house to Joanne's house, which is 12 miles.
>
> c is the distance from Seth's house to Joanne's house, which is unknown.

Substitute these values in the equation to solve the problem.
$9^2 + (12)^2 = c^2$
Square both nine and twelve.
$9^2 = 81$ and $(12)^2 = 144$
Substitute these values in the original equation.
$81 + 144 = c^2$
$$225 = c^2$$

Take the square root of both sides of this equation. The answer is 15.
It is 15 miles from Seth's house to Joanne's house.

BRAIN TICKLERS—THE ANSWERS
Set # 42, page 161

1. $\sqrt{9} = 3, -3$
2. $\sqrt{25} = 5, -5$
3. $\sqrt{36} = 6, -6$
4. $\sqrt{81} = 9, -9$
5. $\sqrt{49} = 7, -7$

6. $\sqrt{100} = 10, -10$
7. $\sqrt{16} = 4, -4$
8. $\sqrt{4} = 2, -2$
9. $\sqrt{64} = 8, -8$
10. $\sqrt{1} = 1, -1$

Set # 43, page 164

1. $\sqrt[3]{27} = 3$
2. $\sqrt[3]{64} = 4$
3. $\sqrt[5]{1} = 1$

4. $\sqrt[4]{16} = 2$
5. $\sqrt[3]{125} = 5$
6. $\sqrt[10]{0} = 0$

Set # 44, page 165

1. If $\sqrt[x]{16} = 4$, then $x = 2$.
2. If $\sqrt[x]{16} = 2$, then $x = 4$.
3. If $\sqrt[x]{25} = 5$, then $x = 2$.

4. If $\sqrt[x]{125} = 5$, then $x = 3$.
5. If $\sqrt[x]{8} = 2$, then $x = 3$.

Set # 45, page 168

1. $\sqrt{-64}$ is undefined.
2. $\sqrt[4]{-16}$ is undefined.
3. $\sqrt[3]{-27} = -3$

4. $\sqrt[5]{-32} = -2$
5. $\sqrt{-49}$ is undefined.
6. $\sqrt[9]{-1} = -1$

Set # 46, page 171

1. $\sqrt{(16)(9)} = 12$

2. $\sqrt{(64)(100)} = 80$

3. $\sqrt{25y^2} = 5y$

4. $\sqrt{(4)(11)} = 2\sqrt{11}$

5. $\sqrt{9y} = 3\sqrt{y}$

Set # 47, page 173

1. $(\sqrt{3})(\sqrt{3}) = 3$

2. $(\sqrt{8})(\sqrt{2}) = 4$

3. $(\sqrt{12})(\sqrt{3}) = 6$

4. $(\sqrt{x})(\sqrt{x}) = x$

5. $(\sqrt{x^3})(\sqrt{x}) = x^2$

Set # 48, page 174

1. $\sqrt{20} = 2\sqrt{5}$

2. $\sqrt{8} = 2\sqrt{2}$

3. $\sqrt{27} = 3\sqrt{3}$

4. $\sqrt{24} = 2\sqrt{6}$

5. $\sqrt{32} = 4\sqrt{2}$

6. $\sqrt{125} = 5\sqrt{5}$

Set # 49, page 179

1. $\dfrac{5}{\sqrt{3}} = \dfrac{5\sqrt{3}}{3}$

2. $\dfrac{10}{\sqrt{2}} = \dfrac{10\sqrt{2}}{2} = 5\sqrt{2}$

3. $\sqrt{\dfrac{16}{3}} = \dfrac{4\sqrt{3}}{3}$

4. $\sqrt{\dfrac{36}{7}} = \dfrac{6\sqrt{7}}{7}$

Set # 50, page 181

1. $7\sqrt{2} + \sqrt{2} = 8\sqrt{2}$

2. $3\sqrt{3} + \sqrt{3} = 4\sqrt{3}$

3. $5\sqrt[4]{x} + 2\sqrt[4]{x} = 7\sqrt[4]{x}$

4. $9\sqrt{5} - \sqrt{5} = 8\sqrt{5}$

5. $2\sqrt{x} - 2\sqrt{x} = 0$

6. $5\sqrt{2x} - 12\sqrt{2x} = -7\sqrt{2x}$

Set # 51, page 186

1. $\sqrt{5} = 5^{\frac{1}{2}}$

2. $\sqrt{x} = x^{\frac{1}{2}}$

3. $\sqrt[3]{12} = 12^{\frac{1}{3}}$

4. $\sqrt[5]{7^2} = 7^{\frac{2}{5}}$

5. $\sqrt[3]{(2x)^2} = (2x)^{\frac{2}{3}}$

Set # 52, page 186

1. $10^{\frac{1}{2}} = \sqrt{10}$

2. $3^{\frac{1}{3}} = \sqrt[3]{3}$

3. $(5xy)^{\frac{1}{2}} = \sqrt{5xy}$

4. $7^{\frac{2}{3}} = \sqrt[3]{7^2}$

5. $(9x)^{\frac{2}{3}} = \sqrt[3]{(9x)^2}$

Super Brain Ticklers, page 187

1. $\sqrt[3]{125} = 5$

2. $\sqrt[3]{-27} = -3$

3. $\sqrt{12} = 2\sqrt{3}$

4. $\sqrt{-49}$ is undefined.

5. $\sqrt{(8)(2)} = 4$

6. $\sqrt{12}\sqrt{3} = 6$

7. $\sqrt{32} = 4\sqrt{2}$

8. $\dfrac{8}{\sqrt{5}} = \dfrac{8\sqrt{5}}{5}$

9. $2\sqrt{3} - 4\sqrt{3} = -2\sqrt{3}$

10. $\left(5\sqrt{2}\right)\left(2\sqrt{18}\right) = 60$

Chapter 9

Quadratic Equations

A *quadratic equation* is an equation with a variable to the second power but no variable higher than the second power. A quadratic equation has the form $ax^2 + bx + c = 0$, where a is not equal to zero. Here are five examples of quadratic equations. In each of these quadratic equations, there is an x^2-term.

$$x^2 + 3x + 5 = 0$$
$$-3x^2 - 4x + 3 = 0$$
$$x^2 + 3x = 0$$

(Notice that this equation does not have a numerical term.)
$$x^2 - 36 = 0$$
(Notice that this equation does not have an x-term.)

The following are *not* quadratic equations.

$x^2 - 4x - 6$ is not a quadratic equation because it does not have an equals sign.

$x^3 - 4x^2 + 2x - 1 = 0$ is not a quadratic equation because there is an x^3-term.

$2x - 6 = 0$ is not a quadratic equation because there is no x^2-term.

A quadratic equation is formed when a linear equation is multiplied by the variable in the equation.

Multiply the linear equation $x + 3 = 0$ by x, and the result is the quadratic equation $x^2 + 3x = 0$.

Multiply $y - 7 = 0$ by $2y$ and the result is $2y^2 - 14y = 0$.

BRAIN TICKLERS   Set # 53

Multiply the following expressions together to form a quadratic equation.

1. $x(3x + 1) = 0$

2. $2x(x - 5) = 5$

3. $-x(2x - 6) = 0$

4. $4x(2x - 3) = 23$

(Answers are on page 243.)

Quadratic equations are also formed when two binomial expressions of the form $x - a$ or $x + a$ are multiplied. The following binomial expressions are quadratic equations.

$$(x - 3)(x - 2) = 0$$
$$(x + 6)(x + 5) = 0$$
$$(2x - 5)(3x + 4) = 0$$

To multiply two binomial terms, complete the following five steps.

Step 1: Multiply the two **First** terms.

Step 2: Multiply the two **Outside** terms.

Step 3: Multiply the two **Inside** terms.

Step 4: Multiply the two **Last** terms.

Step 5: Add the terms and simplify.

Watch how these two expressions are multiplied.

$(x - 3)(x + 2) = 0$

Step 1: Multiply the two **First** terms.
$$(x - 3)(x + 2) = 0$$
The two first terms are x and x.
$$(x)(x) = x^2$$

Step 2: Multiply the two **Outside** terms.
$$(x - 3)(x + 2) = 0$$
The two outside terms are x and 2.
$$(2)(x) = 2x$$

Step 3: Multiply the two **Inside** terms.
$$(x - 3)(x + 2) = 0$$
The two inside terms are -3 and x.
$$(-3)(x) = -3x$$

Step 4: Multiply the two **Last** terms.
$$(x - 3)(x + 2) = 0$$
The two last terms are (-3) and 2.
$$(-3)(2) = -6$$

Step 5: Add the terms and simplify.

First, the terms are added.
$$x^2 + 2x - 3x - 6 = 0$$
Next, the expression is simplified.
$$x^2 - x - 6 = 0$$
Answer: $(x - 3)(x + 2) = x^2 - x - 6 = 0$

Watch how these two binomial expressions are multiplied.

$(x - 5)(x + 5) = 0$

Step 1: Multiply the two **First** terms.
$$(x - 5)(x + 5) = 0$$
The two first terms are x and x.
$$(x)(x) = x^2$$

Step 2: Multiply the two **Outside** terms.
$$(x - 5)(x + 5) = 0$$
The two outside terms are x and 5.
$$(5)(x) = 5x$$

Step 3: Multiply the two **Inside** terms.
$$(x - 5)(x + 5) = 0$$
The two inside terms are -5 and x.
$$(-5)(x) = -5x$$

Step 4: Multiply the two **Last** terms.
$$(x - 5)(x + 5) = 0$$
The two last terms are -5 and 5.
$$(-5)(5) = -25$$

Step 5: Add the terms and simplify.

First, add the terms.
$$x^2 + 5x - 5x - 25 = 0$$
Simplify this expression.
$$x^2 - 25 = 0$$
Answer: $(x - 5)(x + 5) = x^2 - 25 = 0$

Watch how these two binomial expressions are multiplied.

$(3x - 5)(2x - 7) = 0$

Step 1: Multiply the two **First** terms.
$$(3x - 5)(2x - 7) = 0$$
The two first terms are $3x$ and $2x$.
$$(3x)(2x) = 6x^2$$

Step 2: Multiply the two **Outside** terms.
$$(3x - 5)(2x - 7) = 0$$
The two outside terms are $3x$ and -7.
$$(3x)(-7) = -21x$$

Step 3: Multiply the two **Inside** terms.
$$(3x - 5)(2x - 7) = 0$$
The two inside terms are -5 and $2x$.
$$(-5)(2x) = -10x$$

Step 4: Multiply the two **Last** terms.
$$(3x - 5)(2x - 7) = 0$$
The two last terms are -5 and -7.
$$(-5)(-7) = 35$$

Step 5: Add the terms and simplify.

Add the terms.
$$6x^2 - 21x - 10x + 35 = 0$$

Simplify.

$$6x^2 - 31x + 35 = 0$$

Answer: $(3x - 5)(2x - 7) = 6x^2 - 31x + 35 = 0$

PAINLESS TIP

Multiply the parts of two binomial expressions in the following order.

FIRST, OUTSIDE, INSIDE, LAST

Remember the order by remembering the word **FOIL**. The word **FOIL** is the first letters of the words **F**irst, **O**utside, **I**nside, **L**ast.

BRAIN TICKLERS   Set # 54

In each of the following, multiply the binomial expressions to form a quadratic equation.

1. $(x + 5)(x + 2) = 0$

2. $(x - 3)(x + 1) = 0$

3. $(2x - 5)(3x + 1) = 0$

4. $(x + 2)(x - 2) = 0$

(Answers are on page 243.)

Solving Quadratic Equations by Factoring

Now that you know how to recognize quadratic equations, how do you solve them? The easiest way to solve most quadratic equations is by factoring. Before factoring a quadratic equation, you must put it in standard form. Standard form is $ax^2 + bx + c = 0$, where a, b, and c can be any real numbers, except that a cannot equal zero.

Here are four examples of quadratic equations in standard form.

$$x^2 + x - 1 = 0$$
$$5x^2 + 3x - 2 = 0$$
$$-2x^2 - 2 = 0 \quad \text{In this equation, } b = 0.$$
$$4x^2 - 2x = 0 \quad \text{In this equation, } c = 0.$$

Here are three examples of quadratic equations *not* in standard form.

$$3x^2 - 7x = 4$$
$$x^2 = 3x + 2$$
$$x^2 = 4$$

If an equation is not in standard form, you can put it in standard form by adding and/or subtracting the same term from both sides of the equation.

Watch as these quadratic equations are put in standard form.

$3x^2 - 5x = 2$
To put this equation in standard form, subtract 2 from both sides of the equation.
$$3x^2 - 5x - 2 = 2 - 2$$
Simplify.
$$3x^2 - 5x - 2 = 0$$

$x^2 = 2x - 1$
To put this equation in standard form, subtract $2x$ from both sides of the equation.
$$x^2 - 2x = 2x - 2x - 1$$
Simplify.
$$x^2 - 2x = -1$$
Now add 1 to both sides of the equation.
$$x^2 - 2x + 1 = -1 + 1$$
Simplify the equation.
$$x^2 - 2x + 1 = 0$$

$4x^2 = 2x$
To put this equation in standard form, subtract $2x$ from both sides of the equation.
$$4x^2 - 2x = 2x - 2x$$
Simplify the equation.
$$4x^2 - 2x = 0$$

BRAIN TICKLERS   Set # 55

Change the following quadratic equations into standard form.

1. $x^2 + 4x = -6$

2. $2x^2 = 3x - 3$

3. $5x^2 = -5x$

4. $7x^2 = 7$

(Answers are on page 243.)

Once an equation is in standard form, you can solve it by factoring. There are three types of quadratic equations in standard form. You will learn how to solve each of them.

Type I: Type I quadratic equations have only two terms. They have the form $ax^2 + c = 0$. Type I quadratic equations have no middle term, so $b = 0$.

Type II: Type II quadratic equations have only two terms. They have the form $ax^2 + bx = 0$. Type II quadratic equations have no last term, so $c = 0$.

Type III: Type III quadratic equations have all three terms. They have the form $ax^2 + bx + c = 0$. In quadratic equations of this type, a is not equal to zero, b is not equal to zero, and c is not equal to zero.

Now let's see how to solve each of these types.

Type I: In Type I quadratic equations, b is equal to zero and the equation has no x term. Examples:

$$x^2 - 36 = 0$$
$$x^2 + 2 = 0$$
$$2x^2 - 18 = 0$$

To solve Type I equations, use the following four steps.

Step 1: Add or subtract to put the x^2-term on one side of the equals sign and the numerical term on the other side of the equals sign.

Step 2: Multiply or divide to eliminate the coefficient in front of the x^2-term.

Step 3: Take the square roots of both sides of the equals sign.

When solving quadratic equations, the symbol $+/-$ tells you to use *both* the positive and the negative square roots of the number.

Step 4: Check your answer.

Watch how the following Type I quadratic equations are solved.

Solve $x^2 - 36 = 0$.

Step 1: Add or subtract to put the x^2-term on one side of the equals sign and the numerical term on the other side of the equals sign.

Add 36 to both sides of the equation.
$$x^2 - 36 + 36 = 0 + 36$$
Simplify.
$$x^2 = 36$$

Step 2: Multiply or divide to eliminate the coefficient in front of the x^2-term.

There is no coefficient in front of the x^2-term. Go to the next step.

Step 3: Take the square roots of both sides of the equation.
$$\sqrt{x^2} = \sqrt{36}$$

The square root of x^2 is x. The square roots of 36 are 6 and -6.
Solution: If $x^2 = 36$, then $x = 6$ or $x = -6$.

Step 4: Check your answer.

Substitute 6 for x in the original equation, $x^2 - 36 = 0$.

$$(6)^2 - 36 = 0$$

Simplify.

$$36 - 36 = 0$$

$0 = 0$ is a true sentence.

Then, $x = 6$ is a solution to the equation $x^2 - 36 = 0$.

Substitute -6 for x in the original equation, $x^2 - 36 = 0$.

$$(-6)^2 - 36 = 0$$

Simplify.

$$36 - 36 = 0$$

$0 = 0$ is a true sentence.

Then, $x = -6$ is a solution to the equation $x^2 - 36 = 0$.

Solve $4x^2 - 8 = 0$.

Step 1: Add or subtract to put the x^2-term on one side of the equals sign and the numerical term on the other side of the equals sign.

Add 8 to both sides of the equation.

$$4x^2 - 8 + 8 = 0 + 8$$

Simplify.

$$4x^2 = 8$$

Step 2: Multiply or divide to eliminate the coefficient in front of the x^2-term.

Divide both sides of the equation by 4.

$$\frac{4x^2}{4} = \frac{8}{4}$$

Simplify.

$$x^2 = 2$$

Step 3: Take the square roots of both sides of the equation.
$$\sqrt{x^2} = \sqrt{2}$$

The square root of x^2 is x. Two is not a perfect square, so $\sqrt{2}$ is $+\sqrt{2}$ and $-\sqrt{2}$.

Solution: If $4x^2 = 8$, then $x = \sqrt{2}$ or $x = -\sqrt{2}$.

Step 4: Check your answer.

Substitute $\sqrt{2}$ for x in the original equation,
$$4x^2 - 8 = 0.$$
Put $\sqrt{2}$ wherever there is an x.
$$4\left(\sqrt{2}\right)^2 - 8 = 0$$
$$4(2) - 8 = 0$$

$8 - 8 = 0$ is a true sentence.
Then, $x = \sqrt{2}$ is a solution for the equation $4x^2 - 8 = 0$.

Substitute $-\sqrt{2}$ for x in the original equation,
$$4x^2 - 8 = 0.$$
Put $-\sqrt{2}$ wherever there is an x.
$$4\left(-\sqrt{2}\right)^2 - 8 = 0$$
$$4(2) - 8 = 0$$
$8 - 8 = 0$ is a true sentence.
Then, $x = -\sqrt{2}$ is also a solution for the equation $4x^2 - 8 = 0$.

BRAIN TICKLERS   Set # 56

Solve the following Type I quadratic equations.

1. $x^2 - 25 = 0$

2. $x^2 - 49 = 0$

3. $3x^2 - 27 = 0$

4. $2x^2 - 32 = 0$

5. $x^2 - 15 = 0$

6. $3x^2 - 20 = 10$

(Answers are on page 243.)

Type II: In Type II quadratic equations, c is equal to zero. When c is equal to zero, the quadratic equation has no numerical term. Examples:

$$x^2 + 10x = 0$$
$$x^2 - 3x = 0$$

To solve Type II equations, use the following four steps.

Step 1: Factor x out of the equation.

Step 2: Set both factors equal to zero.

Step 3: Solve both equations.

Step 4: Check your answer.

Watch how the following Type II quadratic equations are solved.

Solve $x^2 - 5x = 0$.

Step 1: Factor x out of the equation.
$$x(x - 5) = 0$$

Step 2: Set both factors equal to zero.
$$x = 0; x - 5 = 0$$

Step 3: Solve both equations.

The equation $x = 0$ is solved.
To solve $x - 5 = 0$, add 5 to both sides of the equation.
$$x - 5 + 5 = 0 + 5$$
Simplify.
$$x = 5$$
Solution: If $x^2 - 5x = 0$, then $x = 5$ or $x = 0$.

Step 4: Check your answer.

Substitute 5 for x in the original equation, $x^2 - 5x = 0$.
$$(5)^2 - 5(5) = 0$$
$$25 - 25 = 0$$
$0 = 0$ is a true sentence.
Then, $x = 5$ is a correct answer.

Now substitute 0 for x in the original equation, $x^2 - 5x = 0$.

$$0^2 - 5(0) = 0$$

$0 = 0$ is a true sentence.

Then, $x = 5$ and $x = 0$ are both solutions to the equation.

Solve $2x^2 - 12x = 0$.

Step 1: Factor x out of the equation.

$$x(2x - 12) = 0$$

Step 2: Set both factors equal to zero.

$$x = 0; \; 2x - 12 = 0$$

Step 3: Solve both equations.

The equation $x = 0$ is already solved.
To solve $2x - 12 = 0$, add 12 to both sides of this equation.

$$2x - 12 + 12 = 0 + 12$$

Simplify.

$$2x = 12$$

Divide both sides by 2.

$$x = 6$$

Solution: If $2x^2 - 12x = 0$, then $x = 6$ and $x = 0$.

Step 4: Check the answers.

Substitute 6 for x in the original equation, $2x^2 - 12x = 0$.

$$2(6)^2 - 12(6) = 0$$
$$2(36) - 12(6) = 0$$
$$72 - 72 = 0$$
$$0 = 0$$

Substitute 0 for x in the original equation, $2x^2 - 12x = 0$.

$$2(0)^2 - 12(0) = 0$$
$$0 = 0$$

Then, $x = 6$ and $x = 0$ are both solutions to the original equation.

BRAIN TICKLERS   Set # 57

Solve the following Type II quadratic equations.

1. $x^2 - 2x = 0$

2. $x^2 + 4x = 0$

3. $2x^2 - 6x = 0$

4. $\frac{1}{2}x^2 + 2x = 0$

(Answers are on page 243.)

Type III: In Type III quadratic equations, a is not equal to zero, b is not equal to zero, and c is not equal to zero. In standard form, a quadratic equation of this type has three terms. Examples:

$$x^2 + 5x + 6 = 0$$
$$x^2 - 2x + 1 = 0$$
$$x^2 - 3x - 4 = 0$$

In order to solve a quadratic equation with three terms, factor it into two binomial expressions. Follow these seven steps.

Step 1: Put the equation in standard form.

Step 2: Draw two sets of parentheses and factor the x^2-term.

Step 3: List the pairs of factors of the numerical term.

Step 4: Find the pair of factors that when multiplied together equal the numerical term and when added together equal the term in front of the x.

Step 5: Set each of the two binomial expressions equal to zero.

Step 6: Solve each of the equations.

Step 7: Check your answer.

Watch how this Type III quadratic equation is solved.

Solve $x^2 + 2x + 1 = 0$.

Step 1: Put the equation in standard form.

This equation is in standard form.
Go on to the next step.

Step 2: Draw two sets of parentheses and factor the x^2-term.
$$x^2 = (x)(x)$$
Place the factors in the parentheses.
$$(x \qquad)(x \qquad) = 0$$

Step 3: List the possible pairs of factors of the numerical term.
The numerical term is 1.
What are the possible factors of 1?
$$(1)(1) = 1 \text{ and } (-1)(-1) = 1$$

Step 4: Find the pair of factors that when multiplied together equal the numerical term and when added together equal the term in front of the x.
$$(1)(1) = 1 \text{ and } (1) + (1) = 2$$
Check your choice by multiplying the two binomial expressions to see if you get the original equation.

Place (1) and (1) in the parentheses.
$$(x + 1)(x + 1) = 0$$
Multiply these binomial expressions. Multiply the first terms, outside terms, inside terms, and last terms (FOIL).
Add the results.
$$(x + 1)(x + 1) = x^2 + 1x + 1x + 1 = 0$$
Simplify.
$$x^2 + 2x + 1 = 0$$
This is the original equation.
Therefore, 1 and 1 are the correct factors.

Step 5: Set each of the two binomial expressions equal to zero.
$$x + 1 = 0 \text{ and } x + 1 = 0$$

Step 6: Solve each of the equations.
The two equations are the same, so you need to solve only one of them. Subtract (-1) from both sides of the equation.
$$x + 1 - 1 = 0 - 1$$
Simplify.
$$x = -1$$

Step 7: Check your answer.
Substitute -1 in the original equation,
$x^2 + 2x + 1 = 0$.
$$(-1)^2 + 2(-1) + (1) = 0$$
$$1 + (-2) + 1 = 0$$
$$0 = 0$$
Then, $x = -1$ is the correct answer.

Watch how another Type III equation is solved by factoring.

Solve $x^2 - 5x + 4 = 0$.

Step 1: Put the equation in standard form.
The equation is in standard form.

Step 2: Draw two sets of parentheses and factor the x^2-term. Place the factors in the parentheses.
$$(x \qquad)(x \qquad) = 0$$

Step 3: List all the factors of the numerical term, 4.
$$(2)(2) = 4$$
$$(-2)(-2) = 4$$
$$(4)(1) = 4$$
$$(-4)(-1) = 4$$

Step 4: Find the pair of factors that when multiplied together equal the numerical term and when added together equal the term in front of the x.
(-4) and (-1) is the correct pair of factors
since $(-4)(-1) = +4$ and $(-4)+(-1) = -5$.
Place the factors -4 and -1 in the parentheses.
$$(x - 4)(x - 1) = 0$$

Multiply the two expressions.
$$x^2 - 4x - 1x + 5 = 0$$
Simplify.
$$x^2 - 5x + 5 = 0$$
This is the original equation.
Therefore, -4 and -1 are the correct factors.

Step 5: Set each of the two binomial expressions equal to zero.
$$x - 4 = 0 \text{ and } x - 1 = 0$$

Step 6: Solve each of the equations.
Solve $x - 4 = 0$.
$$x = 4$$
Now solve $x - 1 = 0$.
$$x = 1$$

Step 7: Check your answers.
Substitute 4 in the original equation,
$x^2 - 5x + 4 = 0$.
$$4^2 - 5(4) + 4 = 0$$
Compute the value of this expression.
$$16 - 20 + 4 = 0$$
$$0 = 0$$
This proves that $x = 4$ is a correct solution to the equation.
Now substitute $x = 1$ in the original equation,
$x^2 - 5x + 4 = 0$.
$$(1)^2 - 5(1) + 4 = 0$$
Compute the value of this expression.
$$1 - 5 + 4 = 0$$
$$0 = 0$$
Then, $x = 1$ is also a solution.
The equation $x^2 - 5x + 4 = 0$ has two solutions,
$x = 4$ and $x = 1$.

Now watch as a third Type III quadratic equation is solved by factoring.

Solve $2x^2 + 7x = -6$.

Step 1: Put the equation in standard form.
$$2x^2 + 7x + 6 = 0$$

Step 2: Draw two sets of parentheses and factor the x^2-term. Place the factors in the parentheses.

Notice that the x^2-term has 2 in front of it. The only way to factor $2x^2$ is $(2x)(x)$. Put these terms in the parentheses.
$$(2x \quad)(x \quad) = 0$$

Step 3: List all the factors of the numerical term, 6.
$$(6)(1) = 6$$
$$(-6)(-1) = 6$$
$$(3)(2) = 6$$
$$(-3)(-2) = 6$$

Step 4: Since the x^2-term has a 2 in front of it, use trial and error to figure out which pair of factors will result in the correct quadratic equation.

Substitute each pair of numerical factors in the parentheses. Check to see if, when the two binomial expressions are multiplied, the result is the original equation.
Substitute 6 and 1 into the parentheses.
$$(2x + 6)(x + 1) = 0$$
Multiply the two expressions.
$$2x^2 + 2x + 6x + 6 = 0$$
Simplify.
$$2x^2 + 8x + 6 = 0$$
This is not the original equation, but before testing the next set of factors, SWITCH the positions of the numbers.
$$(2x + 1)(x + 6) = 0$$
Multiply the two binomial expressions.
$$2x^2 + 12x + 1x + 6 = 0$$
Simplify.
$$2x^2 + 13x + 6 = 0$$

By reversing the positions of 6 and 1, a new equation was formed. But this is not the original equation. Put the next pair of factors in the equation.

Substitute -6 and -1 in the parentheses.
$$(2x - 6)(x - 1) = 0$$
Multiply the two binomial expressions to see if you get the original equation.
$$2x^2 - 2x - 6x + 6 = 0$$
Simplify:
$$2x^2 - 8x + 6 = 0$$
This is not the original equation, but before testing the next set of factors, SWITCH the positions of the numbers.
$$(2x - 1)(x - 6) = 0$$
Multiply the two binomial expressions.
$$2x^2 - 12x - 1x + 6 = 0$$
Simplify.
$$2x^2 - 13x + 6 = 0$$
By reversing the positions of -6 and -1, a new equation was formed. But this is not the original equation. Put the next pair of factors in the equation.

Substitute 3 and 2 in the parentheses.
$$(2x + 3)(x + 2) = 0$$
Multiply the two binomial expressions to see if you get the original equation.
$$2x^2 + 4x + 3x + 6 = 0$$
Simplify.
$$2x^2 + 7x + 6 = 0$$
This is the original equation.

Step 5: Set each of the two binomial expressions equal to zero.
$$2x + 3 = 0 \text{ and } x + 2 = 0$$

Step 6: Solve each of the equations.
Solve $2x + 3 = 0$.
Subtract 3 from both sides of the equation.
$$2x + 3 - 3 = 0 - 3$$
$$2x = -3$$

Divide both sides by 2.

$$\frac{2x}{2} = -\frac{3}{2}$$

$$x = -\frac{3}{2}$$

If $2x + 3 = 0$, then $x = -\frac{3}{2}$.

Solve $x + 2 = 0$.

$$x = -2$$

If $x + 2 = 0$, then $x = -2$.

Solution: If $2x^2 + 7x + 6 = 0$, then $x = -\frac{3}{2}$ and $x = -2$.

Step 7: Check your answers.

Substitute $x = -\frac{3}{2}$ and $x = -2$ in the original equation, $2x^2 + 7x + 6 = 0$.

First substitute $x = -\frac{3}{2}$ in $2x^2 + 7x + 6$.

$$2\left(-\frac{3}{2}\right)^2 + 7\left(-\frac{3}{2}\right) + 6 = 0$$

Square $-\frac{3}{2}$.

$$2\left(\frac{9}{4}\right) + 7\left(-\frac{3}{2}\right) + 6 = 0$$

Multiply.

$$\frac{9}{2} - \frac{21}{2} + 6 = 0$$

Change 6 to $\frac{12}{2}$ and add.

$$\frac{9}{2} - \frac{21}{2} + \frac{12}{2} = 0$$

$$0 = 0$$

Then, $x = -\frac{3}{2}$ is a solution to the equation $2x^2 + 7x + 6 = 0$.

Now substitute $x = -2$ in the original equation, $2x^2 + 7x + 6 = 0$.

$$2(-2)^2 + 7(-2) + 6 = 0$$

Square (-2).

$$2(4) + 7(-2) + 6 = 0$$
$$8 + (-14) + 6 = 0$$
$$0 = 0$$

Then, $x = -2$ is a solution to the equation $2x^2 + 7x + 6 = 0$.

BRAIN TICKLERS   Set # 58

Solve the following equations by factoring.

1. $x^2 + 10x + 24 = 0$

2. $x^2 + x - 12 = 0$

3. $2x^2 - 7x + 5 = 0$

4. $x^2 - 2x - 3 = 0$

(Answers are on page 244.)

Here is one last example to try.

Solve $x^2 + x = -1$.

Step 1: Put the equation in standard form.

Add 1 to both sides of the equation.
$$x^2 + x + 1 = -1 + 1$$
Simplify.
$$x^2 + x + 1 = 0$$

Step 2: Factor the x^2-term.

There is only one way to factor the x^2-term.
$$(x)(x) = x^2$$
Place these factors in a set of parentheses.
$$(x \quad)(x \quad) = 0$$

Step 3: List the factors of the numerical term.
$$(1)(1) = 1$$
$$(-1)(-1) = 1$$

Step 4: Try substituting each pair of factors in the parentheses. Check to see if, when the two binomial expressions are multiplied together, the result is the original equation, $x^2 + x = -1$.

Substitute 1 and 1 into $(x\ \ )(x\ \ ) = 0$.
$$(x + 1)(x + 1) = 0$$
Compute the value of this expression.
$$x^2 + 1x + 1x + 1 = 0$$
Simplify.
$$x^2 + 2x + 1 = 0$$
This is not the original equation.

Try the other set of factors.
Substitute -1 and -1 into the parentheses.
$$(x - 1)(x - 1) = 0$$
Compute the value of this expression.
$$x^2 - 1x - 1x + 1 = 0$$
Simplify this expression.
$$x^2 - 2x + 1 = 0$$
This is not the original equation either.

Some quadratic equations cannot be factored, or they are difficult to factor. In order to solve these equations, the quadratic formula was invented. The quadratic formula will be discussed in more detail on page 218.

Solving Quadratic Equations by Completing the Square

One method for solving quadratic equations is completing the square. When completing the square, you create an equation where the left side of the equation is a perfect square and the right side of the equation is a number. You find the solution by finding the square root of both sides of the equation.

A quadratic equation has the form $ax^2 + bx + c = 0$. When completing the square, the equation is changed to $(a + d)^2 = e$. During the process of completing the square, you will compute d and e. Let's get started.

Solve quadratic equations by completing the square in these six *painless* steps.

Step 1: Put the quadratic equation in the form $ax^2 + bx + c = 0$.

Step 2: Make the coefficient of x^2 one by dividing all the terms of the equation by a.

Step 3: Move the term $\frac{c}{a}$ to the right side of the equation.

Step 4: Complete the square on the left side of the equation. To keep the equation balanced, add the same number you added to the left side of the equation to the right side of the equation.

Step 5: Take the square root of both sides of the equation.

Step 6: Solve the linear equation.

Step 7: Check your work.

Watch as this quadratic equation is solved using the complete the square method.

$2x^2 = -12x + 14$

Step 1: Put the quadratic equation in the form $ax^2 + bx + c = 0$.
$$2x^2 = -12x + 14$$
Move everything to the left side of the equation.
$$2x^2 + 12x - 14 = 0$$
$$a = 2, b = 12, \text{ and } c = 14$$

Step 2: Make the coefficient of x^2 one by dividing all the terms in the equation by a.

In the equation $2x^2 + 12x - 14 = 0$, $a = 2$, so divide all the terms of the equation by 2.
$$\frac{2x^2}{2} + \frac{12x}{2} - \frac{14}{2} = 0$$
$$x^2 + 6x - 7 = 0$$

Step 3: Move the c term to the right side of the equation.
In the new equation $a = 1, b = 6$, and $c = -7$.

$$x^2 + 6x - 7 = 0$$

To move 7 to the right side of the equation, just add seven to both sides of the equation.

$$x^2 + 6x - 7 + 7 = 0 + 7$$
$$x^2 + 6x = 7$$

Step 4: Complete the square on the left side of the equation by adding or subtracting a number. Add the same number you added to the left side of the equation to the right side of the equation.

The left side of the equation should have the form $(x + b)^2$.

To make $x^2 + 6x$ a perfect square add 9 to create the expression $x^2 + 6x + 9$, which is $(x + 3)^2$.

If you add 9 to one side of the equation, you have to add 9 to the other side of the equation.

$$x^2 + 6x + 9 = 7 + 9$$

The result is $(x + 3)^2 = 16$.

Step 5: Take the square root of both sides of the equation.

$$\sqrt{(x+3)^2} = \sqrt{16}$$

Since you are taking the square root of 16, the result can be a positive or negative number.

$$x + 3 = 4$$
$$x + 3 = -4$$

Step 6: Solve the linear equations.
If $x + 3 = 4$, then $x = 1$.
If $x + 3 = -4$, then $x = -7$.

Step 7: Check your work.

Substitute $x = 1$ into the original equation
$2x^2 = -12x + 14$.

$$2(1)^2 = -12(1) + 14$$
$$2 = -12 + 14$$
$$2 = 2$$

The solution $x = 1$ is correct.

Substitute $x = -7$ into the original equation
$2x^2 = -12x + 14$.

$$2(-7)^2 = -12(-7) + 14$$
$$98 = 84 + 14$$

The solution $x = -7$ is correct.

Now watch as this quadratic equation is solved by completing the square.

$$x^2 - 2x - 15 = 0$$

Step 1: Start with the quadratic equation in the form
$ax^2 + bx + c = 0$.

The equation is in the correct form.
$a = 1, b = -2$, and $c = -15$

Step 2: Make the coefficient of x^2 one by dividing all the terms in the equation by a.

The coefficient a is equal to one so you can skip this step.

Step 3: Move the term c to the right side of the equation.
Just add 15 to both sides of the equation.

$$x^2 - 2x - 15 + 15 = 15$$
$$x^2 - 2x = 15$$

Step 4: Complete the square on the left side of the equation. Add the same number you added to the left side of the equation to the right side of the equation.

To complete the square add one to each side of the equation.

$$x^2 - 2x + 1 = 15 + 1$$
$$(x - 1)^2 = 16$$

Step 5: Take the square root of both sides of the equation.

Since the square root of 16 is $+4$ and -4, the result is two equations.
$$x - 1 = 4$$
$$x - 1 = -4$$

Step 6: Solve the linear equations.

If $x - 1 = 4$, then $x = 5$.
If $x - 1 = -4$, then $x = -3$.

Step 7: Check your work.

Substitute 5 in the original equation.
$$5^2 - 2(5) - 15 = 0$$
$$25 - 10 - 15 = 0$$
$$0 = 0$$

Five is a correct solution.
Substitute -3 in the original equation.
$$3^2 - 2(-3) - 15 = 0$$
$$9 + 6 - 15 = 0$$
$$0 = 0$$

Negative three is also a correct solution.

 **BRAIN TICKLERS   Set # 59**

Please solve the following quadratic equations by completing the square.

1. $x^2 + 2x - 24 = 0$

2. $3x^2 - 12x = 36$

3. $x^2 + 10x + 16 = 0$

4. $2x^2 - 4x - 6 = 0$

(Answers are on page 244.)

The Quadratic Formula

Another way to solve a quadratic equation is to use the quadratic formula.

When you use the quadratic formula, you can solve quadratic equations without factoring. Just put the equation in standard form, $ax^2 + bx + c = 0$. Substitute the values for a, b, and c in this equation and simplify.

$$x = \frac{-b \pm \sqrt{b^2 - 4ac}}{2a}$$

1+2=3 MATH TALK!

Watch how the quadratic formula is read in Plain English.

$$\frac{-b \pm \sqrt{b^2 - 4ac}}{2a}$$

Negative b plus or minus the square root of the quantity b squared minus four times a times c, all divided by two times a.

When you solve a quadratic equation with the quadratic formula, you may get no answer, one answer, or two answers. The method may look complicated, but it's *painless*. Just follow these four steps.

Step 1: Put the quadratic equation in standard form, $ax^2 + bx + c = 0$.

Step 2: Figure out the values of a, b, and c.

Step 3: Substitute the values of a, b, and c in the quadratic formula and solve for x.

$$x = \frac{-b \pm \sqrt{b^2 - 4ac}}{2a}$$

Step 4: Check your answer.

Watch as an equation is solved using the quadratic formula.

Solve $x^2 + 2x + 1 = 0$.

Step 1: Put the quadratic equation in standard form, $ax^2 + bx + c = 0$.

The equation $x^2 + 2x + 1 = 0$ is in standard form. Go to the next step.

Step 2: Figure out the values of a, b, and c.

The coefficients are $a = 1$, $b = 2$, and $c = 1$.

Step 3: Substitute a, b, and c in the quadratic formula.
$$\frac{-b \pm \sqrt{b^2 - 4ac}}{2a} = \frac{-2 \pm \sqrt{2^2 - 4(1)(1)}}{2(1)}$$

Solve.
$$x = \frac{-2 \pm \sqrt{4 - 4}}{2}$$

Substitute zero for the expression under the radical sign.
$$x = \frac{-2 \pm 0}{2}$$

Divide.
$$x = \frac{-2}{2} = -1$$

If $x^2 + 2x + 1 = 0$, then $x = -1$

Step 4: Check your answer.

Substitute -1 in the original equation.
If the result is a true sentence, then -1 is the correct answer.
Substitute -1 in $x^2 + 2x + 1 = 0$.
$$(-1)^2 + 2(-1) + 1 = 0$$
Compute.
$$1 + (-2) + 1 = 0$$
$$0 = 0$$
Then, $x = -1$ is the correct answer.

Watch as another equation is solved using the quadratic formula.

Solve $2x^2 + 5x + 2 = 0$.

Step 1: Put the equation in standard form.
The equation is in standard form.

Step 2: Figure out the values of a, b, and c.
The coefficients are $a = 2$, $b = 5$, and $c = 2$.

Step 3: Substitute the values of a, b, and c in the quadratic formula.

$$\frac{-b \pm \sqrt{b^2 - 4ac}}{2a} = \frac{-5 \pm \sqrt{(5)^2 - 4(2)(2)}}{2(2)}$$

Compute the value of the numbers under the radical sign.

$$x = \frac{-5 \pm \sqrt{25 - 16}}{4}$$

$$x = \frac{-5 \pm 3}{4}$$

The numerator of this expression, -5 ± 3, is read as "negative five plus or minus three." This expression is equal to two separate expressions.

$$\frac{-5 + 3}{4} \quad \text{and} \quad \frac{-5 - 3}{4}$$

Now the two expressions read as
negative five *plus* three, all divided by four;
negative five *minus* three, all divided by four.

Compute the values of these two expressions.

$$\frac{-5 + 3}{4} = \frac{-2}{4} = -\frac{1}{2}$$

$$\frac{-5 - 3}{4} = \frac{-8}{4} = -2$$

The two possible solutions are $-\frac{1}{2}$ and -2.

To check if $-\frac{1}{2}$ and/or -2 are (is) correct, substitute them in the original equation. If either result is a true sentence, then the corresponding answer is correct.

Step 4: Check your answer.

Start by substituting $-\frac{1}{2}$ in the original equation, $2x^2 + 5x + 2 = 0$.

$$2\left(-\frac{1}{2}\right)^2 + 5\left(-\frac{1}{2}\right) + 2 = 0$$

Compute.

$$2\left(\frac{1}{4}\right) + 5\left(-\frac{1}{2}\right) + 2 = 0$$

Now multiply.

$$\frac{2}{4} + \left(-\frac{5}{2}\right) + 2 = 0$$

Simplify.

$$\frac{1}{2} + \left(-\frac{5}{2}\right) + 2 = 0$$

Compute.

$$-\frac{4}{2} + 2 = 0$$
$$0 = 0$$

This is a true sentence, so $-\frac{1}{2}$ is a correct solution.

Now substitute -2 in the original equation, $2x^2 + 5x + 2 = 0$.

$$2(-2)^2 + 5(-2) + 2 = 0$$

Simplify.

$$2(4) + 5(-2) + 2 = 0$$

Simplify.

$$8 + (-10) + 2 = 0$$
$$0 = 0$$

Therefore, -2 is a correct answer.

Sometimes a solution to a quadratic equation is also called the *root*. In this problem, $-\frac{1}{2}$ and -2 are the *roots* of $2x^2 + 5x + 2 = 0$.

PAINLESS TIP

The expression "±" is read as "plus or minus." When you see this expression, you will get two answers.

$$5 \pm 3 \text{ is } 5 + 3 \text{ or } 5 - 3$$
$$5 + 3 = 8; \ 5 - 3 = 2$$

The two answers to 5 ± 3 are 8 and 2.

BRAIN TICKLERS   Set # 60

Solve the following quadratic equations using the quadratic formula.

1. $x^2 + 4x + 3 = 0$
2. $x^2 - x - 2 = 0$
3. $x^2 - 3x + 2 = 0$
4. $4x^2 - 4x + 1 = 0$
5. $x^2 = 36$

(Answers are on page 244.)

Solving a System of Linear and Quadratic Equations Algebraically

What if you have two equations of which one is a linear equation and the other is a quadratic equation? How do you solve them? It's *painless*; just follow these six steps.

Step 1: Solve for y in the linear equation.

Step 2: Solve for y in the quadratic equation.

Step 3: Set the two equations equal to each other. They are both equal to y, so they are equal to each other.

Step 4: Solve for x in the new equation.

Step 5: Solve for y by substituting the results for x in the linear equation.

Step 6: Substitute the answers in the quadratic equation to check them.

> **REMINDER**
>
> Remember a linear equation has the form $ax + by + c = 0$. The highest degree (or exponent) of either variable is one. In a quadratic equation, one of the variables has a degree of two. A quadratic equation looks like this: $ax^2 + bx + c = 0$.

Sound complicated? Watch as this system of equations is solved using these six *painless* steps.

$$4x + 2y = 12$$
$$y = x^2 + 2x + 1$$

Step 1: Solve for y in the linear equation.
$$4x + 2y = 12$$
$$2y = -4x + 12$$
$$y = -2x + 6$$

Step 2: Solve for y in the quadratic equation.
$$y = x^2 + 2x + 1$$
It's already in terms of y, so you don't have to do anything.

Step 3: Set the two equations equal to each other. Since they are both equal to y, they are equal to each other.
$$-2x + 6 = x^2 + 2x + 1$$

Step 4: Solve.

$$-2x + 6 = x^2 + 2x + 1$$
$$x^2 + 2x + 2x - 6 + 1 = 0$$
$$x^2 + 4x - 5 = 0$$
$$(x + 5)(x - 1) = 0$$
$$x = -5, x = 1$$

Step 5: Solve for y by substituting the results for x in the original linear equation.
The original linear equation is $4x + 2y = 12$.
If $x = -5$, then $4(-5) + 2y = 12$.

$$-20 + 2y = 12$$
$$2y = 32$$
$$y = 16$$

If $x = -5, y = 16$.
If $x = 1$, then $4(1) + 2y = 12$.
$$2y = 12 - 4$$
$$2y = 8$$
$$y = 4$$
If $x = 1, y = 4$.

Step 6: Substitute the answers in the quadratic equation to check them.

The quadratic equation is $y = x^2 + 2x + 1$.
Substitute the solution $x = -5$ and $y = 16$ in the quadratic equation to check the solution.
$$16 = (-5)^2 + 2(-5) + 1$$
$$16 = 25 - 10 + 1$$
$$16 = 16$$
$x = -5$ and $y = 16$ is a correct solution.

Substitute the solution $x = 1$ and $y = 4$ in the quadratic equation to check the solution.
$$4 = (1)^2 + 2(1) + 1$$
$$4 = 1 + 2 + 1$$
$$4 = 4$$

$x = 1$ and $y = 4$ is a correct solution.

Now watch as this system of equations is solved.

$$x + y = 4$$
$$y = x^2 + 2x + 4$$

Step 1: Solve for y in the linear equation.

$$y = 4 - x$$

Step 2: Solve for y in the quadratic equation.

$$y = x^2 + 2x + 4$$

Step 3: Set the two equations equal to each other.

$$4 - x = x^2 + 2x + 4$$

Step 4: Solve.

$$x^2 + 2x + x + 4 - 4 = 0$$
$$x^2 + 3x = 0$$
$$x(x + 3) = 0$$
$$x = 0 \text{ and } x = -3$$

Step 5: Solve for y by substituting the results for x in the linear equation.

Substitute 0 for x in the equation $x + y = 4$. If $x = 0$, $y = 4$.
Substitute -3 for x in the equation $x + y = 4$. If $x = -3$, $y = 7$.

Step 6: Substitute the answers into the quadratic equation to check them.

Substitute $x = 0$ and $y = 4$ in the equation, $y = x^2 + 2x + 4$.
$4 = 0^2 + 2(0) + 4$
$4 = 4$ so the answer $x = 0$ and $y = 4$ is correct.
Substitute $x = -3$ and $y = 7$ in the equation $y = x^2 + 2x + 4$.
$7 = (-3)^2 + 2(-3) + 4$
$7 = 7$ so the answer $x = -3$ and $y = 7$ is correct.

Sometimes there is only one solution. Watch as this pair of equations is solved.

$y = x^2 + x + 1$
$y = x + 1$

Step 1: Solve for y in the linear equation.

The linear equation $y = x + 1$ is in terms of y.

Step 2: Solve for y in the quadratic equation.

The quadratic equation $y = x^2 + x + 1$ is already in terms of y.

Step 3: Set the two equations equal to each other. They are both equal to y, so they are both equal to each other.

Step 4: Solve the resulting equation.

$$x + 1 = x^2 + x + 1$$
$$x^2 + x - x + 1 - 1 = 0$$
$$x^2 = 0$$
$$x = 0$$

Step 5: Solve for y by substituting x in the linear equation.

Substitute 0 for x in the equation $y = x + 1$.
$y = 0 + 1$
$y = 1$

Step 6: Substitute the answers in the quadratic equation to check them.

Substitute $x = 0$ and $y = 1$ in the equation,
$y = x^2 + x + 1$.
$1 = (0)^2 + 0 + 1$
$1 = 1$
$x = 0$ and $y = 1$ is the only solution.

BRAIN TICKLERS   Set # 61

Solve.

1. $y = x^2 + x + 1$
    $y = x + 2$

2. $y = x^2 + 6x + 9$
    $y = x + 3$

3. $y = x^2 + 5x + 9$
    $y = x + 5$

4. $y = x^2 + 2x + 1$
    $y = -x + 5$

(Answers are on page 244.)

Graphing Quadratic Equations

The standard form of a quadratic equation is $y = ax^2 + bx + c$. If you graph this quadratic equation, it will be a parabola. Think of a parabola as a U-shaped curve with specific properties. Some of the U's open up, and some of the U's open down. Some of the U's are fatter, and some of them are skinnier. Whether a parabola opens up or down, is fat or skinny, a parabola expands outward.

Every parabola has a highest or lowest point called the *vertex*. The vertex is the minimum or maximum point of the parabola.

Here are illustrations of three different parabolas.

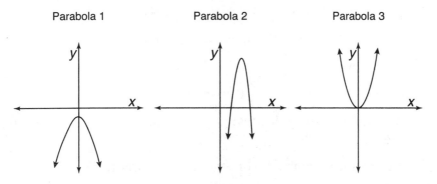

Parabola 1   Parabola 2   Parabola 3

Parabolas 1 and 2 open down, while Parabola 3 opens up. Parabola 1 is the fattest, while Parabola 2 is the skinniest. Parabola 1 has a maximum point since it opens down and the vertex is on the y-axis. The vertex of Parabola 2 is on neither the x- nor the y-axis. The vertex of Parabola 3 is at the origin. The minimum point of Parabola 3 is $(0, 0)$.

You can tell a lot about the graph of a parabola just by looking at the formula of the parabola.

Just look at the a coefficient in front of the x^2-term in the formula $y = ax^2 + bx + c$.

If a is a positive number, the parabola will open up.

If a is a negative number, the parabola will open down.

If a is 0, the equation will not be a parabola. It will be a straight line.

The coefficient a also tells you if the parabola is fat or skinny.

If a is greater than one, then the larger a is the skinnier the parabola will be.

If a is less than negative one, then the smaller a is the skinnier the parabola will be.

If a is between positive one and negative one, the closer the coefficient is to zero, the fatter the parabola will be.

The simplest quadratic equation is $y = x^2$.

In this equation $a = 1, b = 0$, and $c = 0$.

This parabola will open up since a is positive one.

The vertex of this parabola will be at the origin.

The quadratic equation $y = -x^2$ is another simple parabola.

In this equation, $a = -1, b = 0$, and $c = 0$.

This parabola will open downward since a equals negative one.

The vertex of this parabola will also be at the origin.

One of the easiest ways to graph a parabola is to find the vertex and four points on the parabola. Graphing a parabola is *painless* if you follow these five steps.

Step 1: Place the equation in standard form, $y = ax^2 + bx + c$, and determine the values of a, b, and c.

Step 2: Find and plot the vertex of the parabola. The vertex is the highest or lowest point of the parabola. It will be defined by a point (x, y). Find the x-term of the vertex by using the formula $x = \dfrac{-b}{2a}$. Find the y-term of the vertex by substituting the x-term into the original equation.

Step 3: Look at the a-term, and determine if the parabola opens up or down. If a is a positive number, the parabola will open up, and the vertex will be the lowest point on the parabola. If a is a negative number, the parabola will open down and the vertex will be the highest point on the parabola.

Step 4: Look at the x-term of the vertex. Pick two consecutive x-terms on either side of the vertex. Find the corresponding y-terms. Use a chart to display your results.

Step 5: Connect the points of the parabola. Make sure to graph it as a smooth curve.
That wasn't so hard, was it?

Watch as this parabola is graphed using these five painless steps.

Graph the parabola $y = 2x^2$.

Step 1: Place the equation in standard form, $y = ax^2 + bx + c$, and determine the values of a, b, and c.

The equation is in standard form. The coefficients are $a = 2$, $b = 0$, and $c = 0$.

Step 2: Find and plot the vertex of the parabola. The vertex is the highest or lowest point of the parabola. It will be defined by a point (x, y). Find the x-term of the vertex by using the formula $x = \dfrac{-b}{2a}$.

The x-term of the vertex is $\frac{-b}{2a} = \frac{-0}{2(2)} = \frac{-0}{4} = 0$.

Find the y-term by substituting x into the original equation, $y = 2x^2$.
$y = 2(0^2) = 0$
The vertex of the parabola is the point $(0, 0)$.
The vertex is at the origin.

Step 3: Look at the a-term. If a is a positive number, the parabola will open up, and the vertex will be the lowest point on the parabola. If a is a negative number, the parabola will open down, and the vertex will be the highest point on the parabola.

The a-term is 2, so the parabola opens up.

Step 4: Look at the x-term of the vertex. Pick two consecutive x-terms on either side of the vertex. Find the corresponding y-values by substituting the value of x in the equation $y = 2x^2$ to find y. Use a chart to display your results. Notice the vertex is in bold.

x	y
-2	8
-1	2
0	**0**
1	2
2	8

Notice $y = 0$ is the minimum value of y, and the values are symmetrical.

Step 5: Plot the points of the parabola and connect them. Make sure to graph it as a smooth curve.

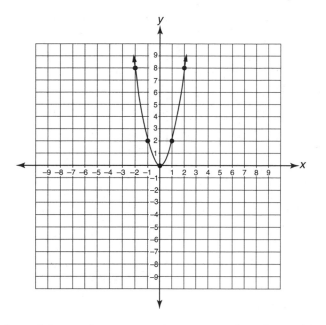

Now watch as this quadratic equation $y = -x^2 - 4x + 4$ is graphed.

Step 1: Place the equation in standard form, $y = ax^2 + bx + c$, and determine the values of a, b, and c.
$y = -x^2 - 4x + 4$ is in standard form.
In this equation, $a = -1$, $b = -4$, and $c = 4$.

Step 2: Find and plot the vertex of the parabola. The vertex is the highest or lowest point of the parabola. It will be defined by a point (x, y). Find the x-term of the vertex by using the formula $x = \dfrac{-b}{2a}$. Find the y term of the point by substituting the x-term into the original equation, $y = -x^2 - 4x + 4$.
The x-term of the vertex is
$\dfrac{-b}{2a} = -\dfrac{-4}{2(-1)} = \dfrac{4}{-2} = -2$
Substitute $x = -2$ into the equation $-x^2 - 4x + 4 = y$ to find the y-term of the vertex.
The y-term of the vertex is
$-(-2)^2 - 4(-2) + 4 = -4 + 8 + 4 = 8$
The vertex is the point $(-2, 8)$.

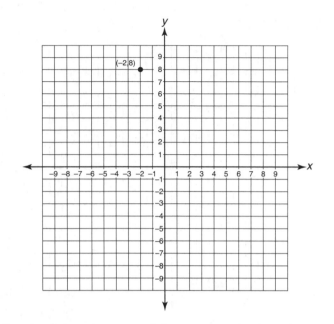

Step 3: Look at the *a*-term, and determine whether the parabola opens up or down. If *a* is a positive number, the parabola will open up, and the vertex will be the lowest point on the parabola. If *a* is a negative number, the parabola will open down, and the vertex will be the highest point on the parabola.

The *a*-term is (-1) so the parabola opens down.

Step 4: Look at the *x*-term of the vertex. Pick two consecutive *x*-terms on either side of the vertex. Find the corresponding *y*-values by substituting the value of *x* in the equation $y = -x^2 - 4x + 4$ to find *y*. Use a chart to display your results. Notice the vertex is in bold.

x	*y*
-4	4
-3	7
-2	**8**
-1	7
0	4

Notice $y = 8$ is the highest point in the chart. It is the maximum value of the parabola. Also notice the *y*-values are symmetrical.

Step 5: Plot the points of the parabola, and connect them. Make sure to graph it as a smooth curve.

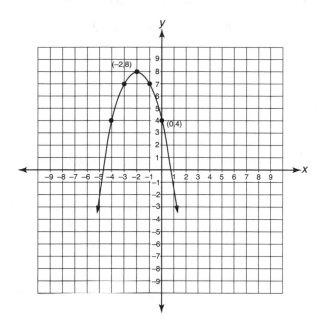

 BRAIN TICKLERS   Set # 62

Find the vertex of each of these parabolas and state whether they open up or down.

1. $y = x^2 - 8x - 9$

2. $y = -x^2 - x + 6$

3. $y = x^2 - 2x - 8$

4. $y = 2x^2 + 8x + 6$

Graph the following parabolas.

5. $y = x^2$

6. $y = -x^2$

(Answers are on page 245.)

Solving a System of Linear and Quadratic Equations Graphically

You can also solve a system of linear and quadratic equations by graphing. It's *painless.*

Just follow these three painless steps:

Step 1: Graph the quadratic equation.

Step 2: On the same graph, graph the linear equation.

Step 3: Find the point or points where the graphs intersect. This is the solution.

The two graphs can intersect two points, one point, or no points. If there are two intersection points, there are two solutions. If there is one intersection point, there is one solution. It is possible that there will be no intersection points and therefore no solutions.

Now watch as this system of two equations is solved by graphing.

$$y = x^2 + 2x + 1$$
$$y = x + 3$$

Step 1: Graph the quadratic equation $y = x^2 + 2x + 1$.

The equation has a single x^2 term. It is a parabola.
This parabola opens up since the x^2 is positive.
To graph a parabola, first find the vertex.
The x-coordinate of the vertex is $\frac{-b}{2a}$.
In this equation, $a = 1$ and $b = 2$.
$\frac{-b}{2a} = \frac{-2}{2(1)} = -1$.
Now find the y-coordinate of the vertex by substituting $x = -1$ in the equation
$y = x^2 + 2x + 1$.
If $x = -1, y = 0$.
The vertex is $(-1, 0)$.
Now that you know the vertex, find five points on the parabola and graph them.

Use $x = -1$ as one point. Use two x-values greater than -1 and two x-values less than one.

x	y
-3	
-2	
-1	
0	
1	

To find y, substitute x into the equation: $x^2 + 2x + 1$.

x	y
-3	4
-2	1
-1	0
0	1
1	4

Now you can graph the parabola.

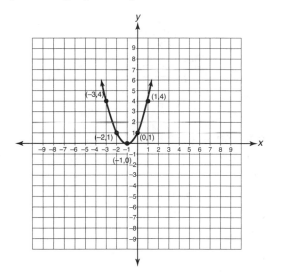

Step 2: Graph the linear equation $y = x + 3$ on the same set of coordinate axes.

To graph a linear equation, just find three points. Substitute $x = -1$, $x = 0$, and $x = 1$ into the equation, $y = x + 3$. Typically, x equals $-1, 0$, and 1 are the easiest points to use.

x	y
-1	2
0	3
1	4

Now graph the points.

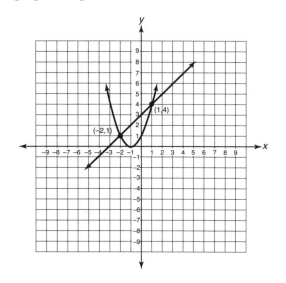

The intersection points are $(1, 4)$ and $(-2, 1)$.
These equations have two intersection points and two solutions.

Now watch as these linear and quadratic equations are graphed on the same set of coordinate axes.

$$y = x^2 + 2x + 1$$
$$y = x$$

Step 1: Graph the quadratic equation $y = x^2 + 2x + 1$.
$y = x^2 + 2x + 1$ is the same equation as in the previous example, so the graph is the same.

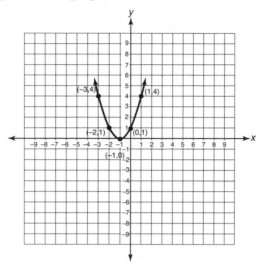

Step 2: Graph the linear equation.
Find three points where $y = x$. Again, use $x = -1$, $x = 0$, and $x = 1$.

x	y
-1	-1
0	0
1	1

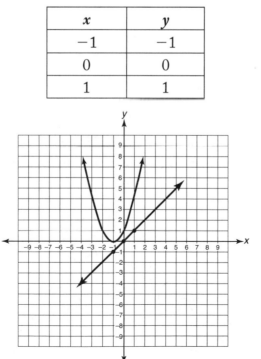

Step 3: Find the points where the graphs intersect.

The graphs do not intersect. There is no solution to this system of equations.

Now watch this final example.

$$y = x^2 + 2x + 1$$
$$y = 0$$

Step 1: Graph the quadratic equation, $y = x^2 + 2x + 1$.

This equation is the same equation used in the previous two examples, so use the same graph.

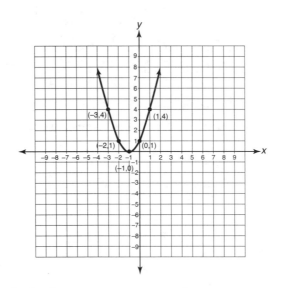

Step 2: Graph the linear equation, $y = 0$.

No matter what number x is, y is always zero.

x	y
-1	0
0	0
1	0
2	0

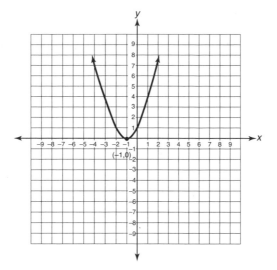

$y = 0$ is the x-axis. There is only one point of intersection between the line and the parabola. The point of intersection is $x = -1$ and $y = 0$.

There is only one solution to the system of equations $y = x^2 + 2x + 1$ and $y = 0$.

The solution is $(-1, 0)$.

PAINLESS TIP

To find the solution of a quadratic equation and a linear equation, graph the equations on the same set of coordinate axes. If there are two points of intersection, there will be two solutions. If there is one point of intersection, there will be one solution. If there are no points of intersection, there will be no solution.

 **BRAIN TICKLERS   Set # 63**

Solve each of the systems of equations by graphing.

1. $y = x^2 - 6x + 2$
    $y = x + 2$

2. $y = -x^2 + 2x$
    $y = x + 5$

(Answers are on page 246.)

Word Problems

Here are two word problems that require factoring to solve. Read each one carefully and watch the *painless* solution.

Problem 1: The product of two consecutive even integers is 48. Find the integers.

First change this problem from Plain English into Math Talk.
The expression "the product" means "multiply."
Let x represent the first even integer.
The second even integer is $x + 2$, because both integers are even.
The word *is* means "=."
48 goes on the other side of the equals sign.
Now the problem can be written as $(x)(x + 2) = 48$.

To solve, multiply this equation.
$x^2 + 2x = 48$

Put the new equation in standard form.
$x^2 + 2x - 48 = 0$
Factor.
$(x + 8)(x - 6) = 0$

Solve for x.
If $x + 8 = 0$, then $x = -8$.
If $x - 6 = 0$, then $x = 6$.

The two consecutive even integers were termed x and "$x + 2$."
If $x = -8$, then $x + 2 = -6$.
If $x = 6$, then $x + 2 = 8$.

Check both pairs of answers.
$(-8)(-6) = 48$; -8 and -6 are a correct solution.
$(6)(8) = 48$; 6 and 8 are a correct solution.

Problem 2: The width of a rectangular swimming pool is 10 feet less than the length of the swimming pool. The surface area of the pool is 600 square feet. What are the length and width of the sides of the pool?

First change the problem from Plain English into Math Talk.
If the length of the pool is l, then the width of the pool is $l - 10$.
The area of any rectangle is length times width, so the area of the pool is $(l)(l - 10)$.
The area of the pool is 600 square feet.
The problem can be written as $(l)(l - 10) = 600$.

To solve, multiply this equation.
$l^2 - 10l = 600$

Put the new equation in standard form.
$l^2 - 10l - 600 = 0$
Factor.
$(l - 30)(l + 20) = 0$

Solve for l.
If $l - 30 = 0$, then $l = 30$.
If $l + 20 = 0$, then $l = -20$.
The length of the pool cannot be a negative number, so the length must be 30 feet. If the length is 30 feet, then the width is $l - 10$, or 20 feet.

Check your answers.
$(30)(20) = 600$. This is correct.
The length of the pool is 30 feet, and the width of the pool is 20 feet.

SUPER BRAIN TICKLERS

Solve for x.

1. $(x + 5)(x - 3) = 0$

2. $x^2 - 3x + 2 = 0$

3. $2x^2 - 3x - 2 = 0$

4. $x(x + 2) = -1$

5. $x^2 - 100 = 0$

6. $2x^2 = 50$

7. $3x^2 - 12x = 0$

8. $3x^2 - 4x = -1$

9. $3(x + 2) = x^2 - 2x$

10. $5(x + 1) = 2(x^2 + 1)$

(Answers are on page 247.)

BRAIN TICKLERS—THE ANSWERS
Set # 53, page 194

1. $3x^2 + x = 0$

2. $2x^2 - 10x = 5$

3. $-2x^2 + 6x = 0$

4. $8x^2 - 12x = 23$

Set # 54, page 197

1. $x^2 + 7x + 10 = 0$

2. $x^2 - 2x - 3 = 0$

3. $6x^2 - 13x - 5 = 0$

4. $x^2 - 4 = 0$

Set # 55, page 199

1. $x^2 + 4x + 6 = 0$

2. $2x^2 - 3x + 3 = 0$

3. $5x^2 + 5x = 0$

4. $7x^2 - 7 = 0$

Set # 56, page 202

1. $x = 5; x = -5$

2. $x = 7; x = -7$

3. $x = 3; x = -3$

4. $x = 4; x = -4$

5. $x = \sqrt{15}; x = -\sqrt{15}$

6. $x = \sqrt{10}; x = -\sqrt{10}$

Set # 57, page 205

1. $x = 2; x = 0$

2. $x = -4; x = 0$

3. $x = 3; x = 0$

4. $x = -4; x = 0$

Set # 58, page 212

1. $(x + 6)(x + 4)$; $x = -6$, $x = -4$

2. $(x + 4)(x - 3)$; $x = 3$, $x = -4$

3. $(2x - 5)(x - 1)$; $x = \frac{5}{2}$, $x = 1$

4. $(x - 3)(x + 1)$; $x = 3$, $x = -1$

Set # 59, page 217

1. $x = 4, x = -6$

2. $x = 6, x = -2$

3. $x = -2, x = -8$

4. $x = 3, x = -1$

Set # 60, page 222

1. $x = -3, x = -1$

2. $x = 2, x = -1$

3. $x = 1, x = 2$

4. $x = \frac{1}{2}$

5. $x = 6, x = -6$

Set # 61, page 227

1. $x = 1$ and $y = 3$, $x = -1$ and $y = 1$

2. $x = -3$ and $y = 0$, $x = -2$ and $y = 1$

3. There is only one solution: $x = -2$ and $y = 3$

4. $x = -4$ and $y = 9$, $x = 1$ and $y = 4$

Set # 62, page 233

1. Vertex $= (4, -25)$, the parabola opens up.

2. Vertex $= \left(-\dfrac{1}{2}, \dfrac{25}{4}\right)$, the parabola opens down.

3. Vertex $= (1, -9)$, the parabola opens up.

4. Vertex $= (-2, -2)$, the parabola opens up.

5.

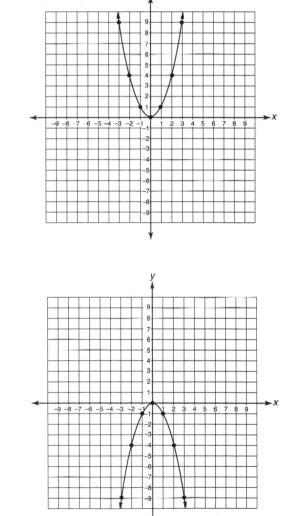

6.

Set # 63, page 239

1. This problem has two solutions: $(0, 2)$ and $(7, 9)$.

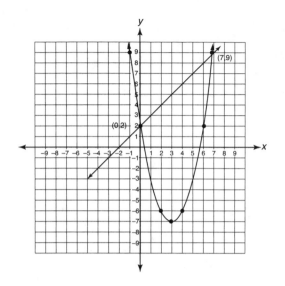

2. This problem has no solution. The line and the parabola do not intersect.

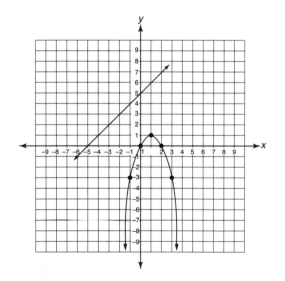

Super Brain Ticklers, page 242

1. $x = -5; x = 3$

2. $x = 1; x = 2$

3. $x = -\dfrac{1}{2}; x = 2$

4. $x = -1$

5. $x = 10; x = -10$

6. $x = 5; x = -5$

7. $x = 0; x = 4$

8. $x = \dfrac{1}{3}; x = 1$

9. $x = 6; x = -1$

10. $x = -\dfrac{1}{2}; x = 3$

Chapter 10

Advanced Topics

Series and Sequences

Did you ever see a group of numbers inside a pair of set brackets? A finite set of numbers in a set of brackets looks like this: $\{1, 8, 5, 3\}$. An infinite set of numbers inside a set of brackets might look something like this: $\{1, 2, 3, 4, \ldots\}$. Sometimes the numbers in the brackets are random numbers, and sometimes the numbers in the brackets have a relationship with each other. Groups of numbers with a specific relationship with each other are sequences.

A sequence is an ordered list of numbers. The sum of the terms in the sequence is a series.

Some sequences have a **definite pattern** that is used to arrive at the sequence's terms.

In some sequences, the terms are in order, and the difference between any term and the next term is constant. You can always figure out the next term of a sequence if you know the terms before it.

There are two types of common sequences, arithmetic and geometric. It's important to understand both of these types of sequences.

Arithmetic sequence

An arithmetic sequence is a list of numbers where the next number in the sequence is always the same amount larger than the previous number in the sequence.

The numbers 2, 4, 6, 8, 10, 12, . . . are an arithmetic sequence. You just add 2 to a number in the sequence to get the next number in the sequence. The number you add to the previous number is called the *common difference.* Two is *the common difference.*

To find the common difference in any sequence, just subtract the first number in the sequence from the second number in the sequence.

In the arithmetic sequence $\{-20, -15, -10, -5, 0, 5, 10, \ldots\}$ the common difference is 5 since $-15 - (-20) = 5$. To find the next number in the sequence, just add 5 to the previous number.

In the arithmetic sequence $\{10, 7, 4, 1, -2, -5, \ldots\}$ the common difference is -3 since $7 - 10 = -3$. To find the next number in the sequence just add (-3) to the previous number in the sequence.

In the arithmetic sequence, $\left\{\frac{1}{4}, \frac{1}{2}, \frac{3}{4}, 1, \frac{5}{4}, \ldots\right\}$ the common

difference is $\frac{1}{4}$ since $\frac{1}{2} - \frac{1}{4} = \frac{1}{4}$.

In the arithmetic sequence $\{1, 0.9, 0.8, 0.7, 0.6, \ldots\}$, the common difference is -0.1 since $0.9 - 1 = -0.1$.

Notice that in an arithmetic sequence the common difference can be positive or negative. It can be a fraction or a decimal.

Once you know both the starting point of a sequence and the common difference of a sequence, you can figure out *any* term of a sequence.

Now that you know what an arithmetic sequence is, it's easy to figure out how to find a specific term in the sequence. To find the *n*th term in an arithmetic sequence, use the following formula: $a_n = a_1 + (n - 1)d$.

⚡ CAUTION—Major Mistake Territory!

The sequence 1, 2, 5, 7, −3, . . . is not an arithmetic sequence. The difference between 1 and 2 is 1, the difference between 2 and 5 is 3, the difference between 5 and 7 is 2, and the difference between 7 and −3 is −10. The numbers in this sequence do not all have the same difference.

1+2=3 MATH TALK!

Let's change $a_n = a_1 + (n − 1)d$ from Math Talk to Plain English.

a_n is the nth term of a sequence where n is a number.

a_1 means the 1st term of the sequence.

For example, a_3 means the 3rd term of the sequence.

For example, a_8 means the 8th term of the sequence.

$n − 1$ means the position of the term in the sequence you are looking for minus 1.

The nth term of an arithmetic sequence is equal to the first term of the sequence plus the quantity n minus one times the common difference.

If you want to find the 5th term in an arithmetic sequence, you add the value of the 1st term to (5 − 1) times the common difference.

Watch as the nth term of some common arithmetic sequences are found.

Find the 8th term of the arithmetic sequence {3, 6, 9, 12, . . .}.

Since this is an easy sequence, we can find the result by expanding the sequence.

The sequence can be rewritten as {3, 6, 9, 12, 15, 18, 21, 24, 27, 30, . . .}.
Count to find the 8th term. It's 24.

Or you can use the formula $a_n = a_1 + (n - 1)d$ to find the 8th term of the sequence. First figure out what all the letters in the formula represent.

a_n is the eighth term.

a_1 is the first term, which is 3.

$(n - 1)$ is the number of the term we are looking for minus one, which is $(8 - 1) = 7$.

d is the common difference, which is 3.

So, by substituting we have $a_n = 3 + 7(3) = 24$.

The eighth term in the sequence $\{3, 6, 9, 12, \ldots\}$ is 24.

Watch as another example is solved.

Find the 101st term in the arithmetic sequence

$$\left\{0, \frac{1}{2}, 1, \frac{3}{2}, \ldots\right\}$$

Since this example asks for the 101st term, it would be too much work to expand the sequence to the 101st term. Using the formula $a_n = a_1 + (n - 1)d$ is much quicker and easier. To solve this problem, first figure out what all the letters in the formula represent.

a_n is the 101st term.

a_1 is the first term, which is 0.

$(n - 1)$ is the number of the term we are looking for minus one, which is $(101 - 1) = 100$.

d is the common difference, which is $\left(\frac{1}{2} - 0\right)$ or $\frac{1}{2}$.

So, by substituting these values into the equation

$a_n = a_1 + (n - 1)d$, we have $a_n = 0 + 100\left(\frac{1}{2}\right) = 50$.

The 101st term in the sequence $\left\{0, \frac{1}{2}, 1, \frac{3}{2}, \ldots\right\}$ is 50.

BRAIN TICKLERS   Set # 64

Answer the following questions about the arithmetic sequence {6, 12, 18, . . .}.

1.   What is the common difference in this arithmetic sequence?

2.   What is the next number in this arithmetc sequence?

3.   What is the 10th term in this sequence?

Answer the following questions for the arithmetic sequence {20, 10, 0, −10, . . .}.

4.   What is the common difference in this arithmetic sequence?

5.   What is the next number in this arithmetic sequence?

6.   What is the 11th term in this arithmetic sequence?

Answer the following questions for arithmetic sequence

$$\left\{ \frac{7}{2}, 4, \frac{9}{2}, 5, \ldots \right\}.$$

7.   What is the common difference in this arithmetic sequence?

8.   What is the next number in this arithmetic sequence?

9.   What is the 21st term in this arithmetic sequence?

(Answers are on page 282.)

Geometric sequence

A second type of sequence is a geometric sequence. In a geometric sequence, instead of adding the same number to every number in the sequence, you multiply every number in the sequence by the same number as long as that number is not zero. In a geometric sequence, the amount you multiply each number in the sequence by to find the next term is called the *common ratio*.

Here are examples of geometric sequences.

$$\{3, 9, 27, 81, \ldots\}$$
$$\{16, 8, 4, 2, \ldots\}$$
$$\{5, -10, 20, \ldots\}$$

To find the common ratio, just divide the second term by the first term.

The common ratio in the geometric sequence $\{3, 9, 27, 81, \ldots\}$ is $\frac{9}{3}$ or 3. Each number in this sequence is three times as large as the number before it. To find the next number in this sequence, just multiply the previous number by 3.

The common ratio in the geometric sequence $\{16, 8, 4, 2, \ldots\}$ is $\frac{8}{16}$ or $\frac{1}{2}$. Each number in this sequence is $\frac{1}{2}$ as large as the number before it. To find the next number in this sequence, just multiply the previous number by $\frac{1}{2}$.

The common ratio in the geometric sequence $\{5, -10, 20, \ldots\}$ is $-\frac{10}{5}$ or -2. To find the next number in this sequence, just multiply the previous number by -2.

> **PAINLESS TIP**
>
> The common ratio in a geometric sequence cannot be zero because 0 times any number is 0 and the result would be a sequence of all zeros.
>
> If the common ratio is zero, the sequence would look like this: $\{n, 0, 0, \ldots\}$, where n is the first number in the sequence. n may or may not be zero.

Finding the *n*th term in a geometric sequence

It's easy to find the next term in a geometric sequence. Just multiply the previous term by the common ratio. But what if you want to find the 10th term or the 100th term or even the 1,000th term? There's a shortcut.

To find any term in a geometric sequence, just use the formula $a_n = a_1 r^{n-1}$, where a_1 is the first term of the sequence, r is the common ratio, and n is the number representing the position of the term you are trying to find.

1+2=3 MATH TALK!

Let's change $a_n = a_1 r^{n-1}$ from Math Talk to Plain English.

a_n means the nth term, which is the term you are looking for.

a_1 is the first term in the sequence.

r is the common ratio.

$n - 1$ is the exponent where n is the term in the sequence you are looking for.

The nth term of a geometric sequence is equal to the first term in the sequence times the common ratio raised to the $n - 1$ power.

If you want to find the 5th term in a geometric sequence, you multiply the value of the 1st term by the common ratio raised to the $(5 - 1)$ or the fourth power.

Examples:

Find the 6th term in the geometric sequence, $\{10, 20, 40, 80, \ldots\}$.
To solve this problem, you just expand the sequence using the common ratio of 2.

Watch: $\{10, 20, 40, 80, 160, 320, 640, \ldots\}$.

The sixth term is 320.

Or you could use the formula $a_1 r^{n-1}$ to find the sixth term in the sequence.

a_1 is the first term, which is 10.

r is the common ratio, which is 2.

$n - 1$ is the 6th term minus 1, which is 5.

Now compute: $a_1 r^{n-1} = 10(2^5) = 10(32) = 320$

Both methods yield the same result.

What is the 8th term in the geometric sequence $\{2, 6, 18, 54, \ldots\}$? Use the formula $a_1 r^{n-1}$ to find the eighth term.

a_1 is the first term, which is 2.

r is the common ratio, which is 3.

$n - 1$ is the 8th term minus one, which is 7.

Now compute: $a_1 r^{n-1} = 2(3^7) = 2(2{,}187) = 4{,}374$

What is the 6th term in the geometric sequence below?

$$\{8, -4, 2, -1, \ldots\}$$

Use the formula $a_1 r^{n-1}$ to find the sixth term.

a_1 is the first term, which is 8.

r is the common ratio. To find the common ratio, divide the second number in the sequence by the first number in the sequence.
$$-\frac{4}{8} = -\frac{1}{2}$$
$n - 1$ is the 6th term minus one, which is 5.

Now compute: $a_1 r^{n-1} = 8\left(-\frac{1}{2}\right)^5 = 8\left(-\frac{1}{32}\right) = -\frac{1}{4}$

BRAIN TICKLERS   Set # 65

Answer the following questions about the geometric sequence {3, 6, 12, . . .}.

1. What is the common ratio in this geometric sequence?

2. What is the next number in this geometric sequence?

3. What is the 7th term in this geometric sequence?

Answer the following questions for the geometric sequence {5, −5, 5, −5, . . .}.

4. What is the common ratio in this geometric sequence?

5. What is the next number in this geometric sequence?

6. What is the 20th term in this geometric sequence?

Answer the following questions for the geometric sequence {256, 128, 64, 32, . . .}.

7. What is the common ratio in this geometric sequence?

8. What is the next number in this geometric sequence?

9. What is the 10th term in this geometric sequence?

(Answers are on page 282.)

Pascal's Triangle and Binomial Coefficients

Pascal's triangle, named after French mathematician Blaise Pascal, is a special triangle created from staggered rows of numbers. Each row of numbers is created from the row of numbers above it. Look at the first ten rows of Pascal's triangle.

```
                1
              1   1
            1   2   1
          1   3   3   1
        1   4   6   4   1
      1   5  10  10   5   1
    1   6  15  20  15   6   1
  1   7  21  35  35  21   7   1
1   8  28  56  70  56  28   8   1
1  9  36  84 126 126  84  36  9   1
```

Here are five interesting characteristics of Pascal's triangle.

1. In Pascal's triangle the number of the row is the same as the number of terms in the row. The first row has one term, the second row has two terms, and the third row has three terms. The eighth row has eight terms and, no surprise, the hundredth row (not pictured here) has one hundred terms.

2. All the numbers on the right and left diagonals are ones.

3. Each number in the interior of the triangle is computed by adding the two numbers above it in the triangle.

> Look at the fourth row of the triangle. It contains the numbers 1, 3, 3, 1. The 1's in this row are part of the exterior diagonals. The 3's are computed by adding 2 + 1 in the row above.

> Look at the seventh row of the triangle. The fourth term is 20. The number 20 was calculated by adding 10 and 10 in the 6th row.

> In the tenth row the 3rd term is 36. Thirty-six was obtained by adding 8 and 28 in the ninth row.

4. Pascal's triangle is symmetrical. If you put a vertical line down the center of the triangle, the numbers on the left side of the triangle match the numbers on the right side of the triangle.

5. When you add the numbers in each row, the sum of the number in each row is twice the sum of the row above it. The sum of each row is also a consecutive power of two.

> Sum of the first row = $1 = 2^0$

> Sum of the second row = $2 = 2^1$

> Sum of the third row = $4 = 2^2$

> Sum of the fourth row = $8 = 2^3$

> Sum of the fifth row = $16 = 2^4$

BRAIN TICKLERS   Set # 66

1. What is the fourth term in the fifth row of Pascal's triangle?

2. What is the third term in the eighth row of Pascal's triangle?

3. What is the first term in the 20th row of Pascal's triangle?

4. What two terms are added together in the 8th row of Pascal's triangle to form 70 in row 9?

5. How many terms will the 20th row of Pascal's triangle have?

6. Create the 11th row of Pascal's triangle.

7. What is the sum of the numbers in the sixth row of Pascal's triangle?

8. What is the sum of the numbers in the 50th row of Pascal's triangle in terms of powers of two?

(Answers are on page 282.)

Binomial expansion and Pascal's triangle

Binomial expansion refers to expanding a binomial number. Watch as the binomial $(a + b)$ is expanded!

$$(a + b)^0 = 1$$
$$(a + b)^1 = a + b$$
$$(a + b)^2 = (a + b)(a + b) = a^2 + 2ab + b^2$$
$$(a + b)^3 = (a + b)(a + b)(a + b) = a^3 + 3a^2b + 3ab^2 + b^3$$

Expanding a binomial expression can be a lot of work, but with the help of Pascal's triangle, it's *painless*. Just follow these three rules.

Rule 1: There is one more term in the expression than the number of the exponent in the expansion.

When the exponent is 0 there is 1 term in the expression.
$(a + b)^0$ has one term.
When the exponent is 1 there are 2 terms in the expression.
$(a + b)^1$ has two terms.
When the exponent is 2 there are 3 terms in the expression.
$(a + b)^2$ has three terms.
When the exponent is 3 there are 4 terms in the expression.
$(a + b)^3$ has four terms.

Rule 2: The exponents of the consecutive terms of each variable follow either an ascending or a descending pattern. The exponents of the first variable start with the same number the binomial is raised to and decreases to zero as the binomial is expanded. The exponents of the second variable start at zero and increase for each consecutive term ending at the same number to which the binomial is raised.

This binomial $(a + b)^2$ is raised to the second power, so two is the highest value for each variable. When expanded, $(a + b)^2 = a^2 + 2ab + b^2$, but if all the hidden variables and exponents were displayed, it would look like this: $a^2b^0 + 2a^1b^1 + a^0b^2$. Look at the variables in this expansion. The exponents of the variable a are 2, 1, and 0 consecutively. The exponents of variable b are 0, 1, and 2 consecutively.

The expansion of $(a + b)^3 = a^3 + 3a^2b + 3ab^2 + b^3$ but if the hidden variables and exponents are shown it becomes $a^3b^0 + 3a^2b^1 + 3a^1b^2 + a^0b^3$. The exponents for a are 3, 2, 1, and 0 consecutively, and the exponents for b are 0, 1, 2, and 3 consecutively.

Rule 3: The coefficients in the expansion of a binomial match the numbers in a row of Pascal's triangle.

$(a + b)^0 = 1$ There is only one coefficient. It is one.
$(a + b)^1 = 1a + 1b$ The coefficients are 1 and 1.
$(a + b)^2 = 1a^2 + 2ab + 1b^2$ The coefficients are 1, 2, and 1.
$(a + b)^3 = 1a^3 + 3a^2b + 3ab^2 + 1b^3$ The coefficients are 1, 3, 3, 1.

The coefficients of these expansions match Pascal's triangle! Just use the row of the triangle that is one greater than the exponent. For example, in the expansion of $(a + b)^4$, use the 5th row of the triangle. For the expansion of $(a + b)^8$, use the 9th row.

$$1$$
$$1 \quad 1$$
$$1 \quad 2 \quad 1$$
$$1 \quad 3 \quad 3 \quad 1$$

Now follow these steps to expand a binomial expression. Since you already know a lot about binomial expressions, it's *painless*.

Step 1: Add a series of the term ab together. The number of ab's to be added is one more than the exponent in the expansion.

Step 2: Place the descending exponents of the a-term starting with the exponent of the binomial expansion.

Step 3: Place the ascending exponents of the b-term on all the b-terms starting with 0.

Step 4: Use the numbers of a single row of Pascal's triangle to determine the coefficients of the terms. The row should be one greater than the power of the expansion.

Step 5: Add all the terms together and simplify.

Study the following binomial expansions using Pascal's triangle.

What is $(a + b)^5$?

Step 1: Add a series of ab terms together. The number of ab's to be added is one more than the exponent in the expansion. Since the expansion is $a + b$ to the fifth power, add six ab terms together.

$$ab + ab + ab + ab + ab + ab$$

Step 2: Place the descending exponents of the a-term on all the a-terms starting with the number of the exponents in the binomial expansion.

$$a^5b + a^4b + a^3b + a^2b + a^1b + a^0b$$

Step 3: Place the ascending exponents of the b-term on all the b-terms starting with 0.

$$a^5b^0 + a^4b^1 + a^3b^2 + a^2b^3 + a^1b^4 + a^0b^5$$

Step 4: Use the numbers of a single row of Pascal's triangle to determine the coefficients of the terms.

Since the expansion is $(a + b)^5$, use the 6th row of Pascal's triangle. The 6th row has 6 terms and will fit the expansion perfectly.

The 6th row of Pascal's triangle is 1 5 10 10 5 1.

Place the numbers as the coefficients in the expression.

$$1a^5b^0 + 5a^4b^1 + 10a^3b^2 + 10a^2b^3 + 5a^1b^4 + 1a^0b^5$$

Step 5: Simplify.

Remove coefficients of one, since they are implied.

$$a^5b^0 + 5a^4b^1 + 10a^3b^2 + 10a^2b^3 + 5a^1b^4 + a^0b^5$$

Remove any variables with an exponent of 0, since any variable to the 0 power is 1.

$$a^5 + 5a^4b^1 + 10a^3b^2 + 10a^2b^3 + 5a^1b^4 + b^5$$

Erase any exponents of 1 but don't erase the variable. If a variable has no exponent showing, an exponent of 1 is implied.

$$a^5 + 5a^4b + 10a^3b^2 + 10a^2b^3 + 5ab^4 + b^5$$

The correct answer is

$$(a + b)^5 = a^5 + 5a^4b + 10a^3b^2 + 10a^2b^3 + 5ab^4 + b^5.$$

Binomial expansions are easy. Here is another example for you to study.

What is $(a + b)^7$?

Step 1: Add a series of ab-terms together. The number of ab's to be added is one more than the exponent in the expansion. Since the expansion is $a + b$ to the seventh power, add eight ab-terms together.

$$ab + ab + ab + ab + ab + ab + ab + ab$$

Step 2: Place the descending exponents of the a-term starting with the number of the exponent in the binomial expansion. End with 0 as the exponent of a.

$$a^7b + a^6b + a^5b + a^4b + a^3b + a^2b + a^1b + a^0b$$

Step 3: Place the ascending exponents of the b-term on all the b-terms starting with 0.

$$a^7b^0 + a^6b^1 + a^5b^2 + a^4b^3 + a^3b^4 + a^2b^5 + a^1b^6 + a^0b^7$$

Step 4: Use the numbers of a single row of Pascal's triangle to determine the coefficients of the terms.

Since the expansion is $(a + b)^7$, use the 8th row of Pascal's triangle. The 8th row has 8 terms and will fit the expansion perfectly.

The 8th row of Pascal's triangle is 1 7 21 35 35 21 7 1. Place the numbers as the coefficients in the expression

$$1a^7b^0 + 7a^6b^1 + 21a^5b^2 + 35a^4b^3 + 35a^3b^4 + 21a^2b^5 + 7a^1b^6 + 1a^0b^7$$

Step 5: Simplify.

Remove any coefficients of one, since they are implied.

$$a^7b^0 + 7a^6b^1 + 21a^5b^2 + 35a^4b^3 + 35a^3b^4 + 21a^2b^5 + 7a^1b^6 + a^0b^7$$

Remove any variables with an exponent of 0, since any variable to the 0 power is 1.

$$a^7 + 7a^6b^1 + 21a^5b^2 + 35a^4b^3 + 35a^3b^4 + 21a^2b^5 + 7a^1b^6 + b^7$$

Remove any exponents of 1. If a variable has no exponent showing, an exponent of 1 is implied.

$$a^7 + 7a^6b + 21a^5b^2 + 35a^4b^3 + 35a^3b^4 + 21a^2b^5 + 7ab^6 + b^7$$

The correct answer is

$$(a + b)^7 = a^7 + 7a^6b + 21a^5b^2 + 35a^4b^3 +$$
$$35a^3b^4 + 21a^2b^5 + 7ab^6 + b^7$$

BRAIN TICKLERS   Set # 67

1. How many terms in the binomial expansion $(a + b)^8$?

2. How many terms in the binomial expansion $(a + b)^{1,000}$?

3. Simplify the term $1a^5b^0$.

4. Expand $(a + b)^4$.

5. Expand $(a + b)^9$.

6. Expand $(a + b)^{11}$.

(Answers are on page 283.)

Matrices

A matrix (plural: matrices) is a set of numbers organized into rows and columns inside a pair of brackets. Some matrices are small, and others are large. The number of rows and the number of columns the matrix has are the dimensions of the matrix. You write the dimensions of a matrix like this: rows × columns. The numbers inside the matrix are called the *elements* of the matrix. The elements in a matrix are labeled by their location in the matrix.

The smallest matrix possible is a 1 × 1 matrix.

$$[4]$$

A matrix with three rows and four columns is a 3 × 4 matrix.

$$\begin{bmatrix} 3 & 5 & 8 & 1 \\ 8 & 6 & 1 & 0 \\ 2 & 7 & 2 & 1 \end{bmatrix}$$

A 4 × 1 matrix has four rows and one column.

$$\begin{bmatrix} 3 \\ 2 \\ 1 \\ 5 \end{bmatrix}$$

1+2=3 MATH TALK!

If a matrix has six rows and one column, it is called a 6 × 1 matrix. This is read as a six by one matrix. Even though the multiplication sign is used, it is not a six times one matrix.

Look at this matrix. This is a 3 × 3 matrix. It has three rows and three columns.

$$\begin{bmatrix} 5 & 2 & 4 \\ 0 & 1 & 6 \\ 7 & 3 & 8 \end{bmatrix}$$

The numbers 0, 1, 2, 3, 4, 5, 6, 7, 8 are the elements of this matrix.

Row 1 consists of the numbers 5, 2, and 4.

Row 2 consists of the numbers 0, 1, and 6.

Row 3 consists of the numbers 7, 3, and 8.

Column 1 consists of the numbers 5, 0, and 7.

Column 2 consists of the numbers 2, 1, and 3.

Column 3 consists of the numbers 4, 6, and 8.

The number 2 is in the first row and second column and is labeled as 2_{12}.

The number 8 is in the third row and third column and is labeled as 8_{33}.

BRAIN TICKLERS   Set # 68

Study this matrix, and answer the questions below.

$$\begin{bmatrix} 2 & 5 \\ 18 & 6 \\ 0.5 & 0 \\ 3 & 1 \end{bmatrix}$$

1. What are the dimensions of this matrix?

2. In what row is the number 18 located?

3. In what column is the number 1 located?

4. What element is located at row 4, column 1?

5. What element is located at row 3, column 1?

(Answers are on page 283.)

Reduced row echelon form

Reduced Row Echelon Form for a square ($n \times n$) matrix is a specific form of a matrix where the leading element of each row is a 1 and the rest of the elements in each column are zeros. Each matrix in Reduced Row Echelon Form has a diagonal of ones.

Here are examples of some matrices in Reduced Row Echelon Form. This is a 2 × 2 matrix.

$$\begin{bmatrix} 1 & 0 \\ 0 & 1 \end{bmatrix}$$

This is a 3 × 3 matrix in Reduced Row Echelon Form.

$$\begin{bmatrix} 1 & 0 & 0 \\ 0 & 1 & 0 \\ 0 & 0 & 1 \end{bmatrix}$$

Finally, this is a 5 × 5 matrix in Reduced Row Echelon Form.

$$\begin{bmatrix} 1 & 0 & 0 & 0 & 0 \\ 0 & 1 & 0 & 0 & 0 \\ 0 & 0 & 1 & 0 & 0 \\ 0 & 0 & 0 & 1 & 0 \\ 0 & 0 & 0 & 0 & 1 \end{bmatrix}$$

Changing a matrix to reduced row echelon form

It is possible to change any matrix to Reduced Row Echelon Form by a process called Gaussian Elimination. It may sound complicated, but trust me, it's *painless*.

Gaussian Elimination is based on three simple rules.

Rule 1: The elements of any row can be interchanged with the elements of another row.

Rule 2: The elements of any row can be multiplied by a nonzero number.

Rule 3: Any row can be changed by adding or subtracting the elements of another row.

To change any 2 × 2 matrix to Reduced Row Echelon Form, follow these four steps. These steps are based on the above rules.

Step 1: Operate on the rows to change the number in Row 1, Column 1 to a 1.

Step 2: Operate on the rows to change the number in Row 2, Column 1 to a 0.

Step 3: Operate on the rows to change the number in Row 2, Column 2 to a 1.

Step 4: Operate on the rows to change the number in Row 1, Column 2 to a 0.

Or just follow the order of the letters in this matrix to tell you which order to operate on the numbers in the matrix.

$$\begin{bmatrix} A & D \\ B & C \end{bmatrix}$$

Now watch as this 2 × 2 matrix is put in Reduced Row Echelon Form.

$$\begin{bmatrix} 1 & 4 \\ 3 & 2 \end{bmatrix}$$

Step 1: Operate on the rows to change the number in Row 1, Column 1 to a 1.

There is already a one in Row 1, Column 1.

$$\begin{bmatrix} 1 & 4 \\ 3 & 2 \end{bmatrix}$$

Step 2: Operate on the rows to change the number in Row 2, Column 1 to a 0.

$-3R_1 + R_2$ means multiply all the elements in Row 1 by a -3 and add the result to Row 2.

$$-3R_1 + R_2 \quad \begin{bmatrix} 1 & 4 \\ 3 & 2 \end{bmatrix} \quad \blacktriangleright \quad \begin{bmatrix} 1 & 4 \\ 0 & -10 \end{bmatrix}$$

Step 3: Operate on the rows to change the number in Row 2, Column 2 to a 1.

$-\dfrac{1}{10} R_2$ means multiply all the elements in Row 2 by $-\dfrac{1}{10}$.

$-\dfrac{1}{10} R_2$ $\begin{bmatrix} 1 & 4 \\ 0 & -10 \end{bmatrix}$ 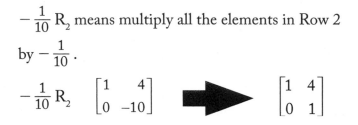 $\begin{bmatrix} 1 & 4 \\ 0 & 1 \end{bmatrix}$

Step 4: Operate on the rows to change the number in Row 1, Column 2 to a 0.

$-4R_2 + R_1$ means multiply Row 2 by -4, and add the result to Row 1.

$-4R_2 + R_1$ $\begin{bmatrix} 1 & 4 \\ 0 & 1 \end{bmatrix}$ 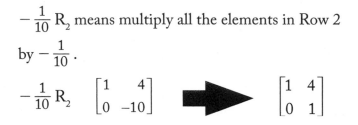 $\begin{bmatrix} 1 & 4 \\ 0 & 1 \end{bmatrix}$

This matrix is now in Reduced Row Echelon Form.
Now watch as this matrix is put in Reduced Row Echelon Form.

$$\begin{bmatrix} 3 & -6 \\ 0 & 5 \end{bmatrix}$$

Step 1: Operate on the rows to change the number in Row 1, Column 1 to a 1.

$\dfrac{1}{3} R_1$ means multiply all the elements in Row 1 by $\dfrac{1}{3}$.

$\dfrac{1}{3} R_1$ $\begin{bmatrix} 3 & -6 \\ 0 & 5 \end{bmatrix}$  $\begin{bmatrix} 1 & -2 \\ 0 & 5 \end{bmatrix}$

Step 2: Operate on the rows to change the number in Row 2, Column 1 to a 0.

It is already a zero. Move on to the next step.

$$\begin{bmatrix} 1 & -2 \\ 0 & 5 \end{bmatrix}$$

Step 3: Operate on the rows to change the number in Row 2, Column 2 to a 1.

$\frac{1}{5} R_2$ means multiplying all the elements in Row 2 by $\frac{1}{5}$.

$$\frac{1}{5} R_2 \quad \begin{bmatrix} 1 & -2 \\ 0 & 5 \end{bmatrix} \quad \blacktriangleright \quad \begin{bmatrix} 1 & -2 \\ 0 & 1 \end{bmatrix}$$

Step 4: Operate on the rows to change the number in Row 1, Column 2 to a 0.

$2R_2 + R_1$ means multiply all the elements in R_2 by 2 and add the result to R_1.

$$2R_2 + R_1 \quad \begin{bmatrix} 1 & -2 \\ 0 & 1 \end{bmatrix} \quad \blacktriangleright \quad \begin{bmatrix} 1 & 4 \\ 0 & 1 \end{bmatrix}$$

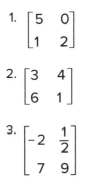

BRAIN TICKLERS   Set # 69

Change the following matrices to Reduced Row Echelon Form.

1. $\begin{bmatrix} 5 & 0 \\ 1 & 2 \end{bmatrix}$

2. $\begin{bmatrix} 3 & 4 \\ 6 & 1 \end{bmatrix}$

3. $\begin{bmatrix} -2 & \frac{1}{2} \\ 7 & 9 \end{bmatrix}$

(Answers are on page 283.)

Solving linear equations with matrices

You can use matrices to solve a system of linear equations.

Step 1: Put the equations in the form $ax + by = c$.

Step 2: Enter the equations into the matrix. Put the coefficients of the variables of the first equation in Row 1 in the matrix.

Place the coefficients of the variables of the second equation in Row 2 of the matrix. Place a vertical line after the coefficients. This vertical line tells you this is an augmented matrix and it separates the variable coefficients from the constants. Put the numerals in the equations (the c-term) in the third column of the matrix.

Step 3: Put the numbers in the matrix to the left of the vertical line in Reduced Row Echelon Form.

Step 4: Place the elements of the matrix back into linear equations and solve.

Watch as this system of equations is solved using matrices.

$x + 3y = 7$
$2x + y = 9$

Step 1: Put the equations in the form $ax + by = c$.

Both equations are in the form $ax + by = c$.

Step 2: Enter the equations into the matrix. Put the coefficients of the variables of the first equation in Row 1 in the matrix. Place the coefficients of the variables of the second equation in Row 2 of the matrix. Place a vertical line after the coefficients. Put the numerals in the equations in the third column of the matrix.

$$\begin{bmatrix} 1 & 3 & | & 7 \\ 2 & 1 & | & 9 \end{bmatrix}$$

Step 3: Put the numbers in the matrix to the left of the vertical line in Reduced Row Echelon Form.

Multiply Row 1 by -2, and add the result to Row 2.

$-2R_1 + R_2$ $\begin{bmatrix} 1 & 3 & | & 7 \\ 2 & 1 & | & 9 \end{bmatrix}$ ➡ $\begin{bmatrix} 1 & 3 & | & 7 \\ 0 & -5 & | & -5 \end{bmatrix}$

Multiply Row 2 by $\frac{3}{5}$ and add to Row 1.

$\frac{3}{5}R_2 + R_1$ $\begin{bmatrix} 1 & 3 & | & 7 \\ 0 & -5 & | & -5 \end{bmatrix}$ ➡ $\begin{bmatrix} 1 & 0 & | & 4 \\ 0 & -5 & | & -5 \end{bmatrix}$

Multiply Row 2 by $-\frac{1}{5}$.

$$-\frac{1}{5}R_2 \quad \begin{bmatrix} 1 & 0 & | & 4 \\ 0 & -5 & | & -5 \end{bmatrix} \quad \blacktriangleright \quad \begin{bmatrix} 1 & 0 & | & 4 \\ 0 & 1 & | & 1 \end{bmatrix}$$

The matrix is now in Reduced Row Echelon Form. Notice the numbers to the right of the vertical line are not in Reduced Row Echelon Form.

Step 4: Place the elements of the matrix back into the linear equations and solve.

$$1x + 0\,y = 4$$
$$0x + 1y = 1$$

Now solve the two equations.

$x = 4$ and $y = 1$

CAUTION—Major Mistake Territory!

Make sure the equation is in standard form, $ax + by = c$, before entering it in the matrix.

The equation $3x - 2y + 5 = 0$ would be entered as a row into the matrix as 3, −2, and −5, NOT 3, −2, 5. You first have to change the equation, $3x - 2y + 5 = 0$, to standard form, which would be $3x - 2y = -5$.

Now watch as one more system of linear equations is solved using matrices.

$$2x + 5y = 0$$
$$x + y + 3 = 0$$

Step 1: Put the equations in the form $ax + by = c$.

$$2x + 5y = 0$$
$$x + y = -3$$

Step 2: Enter all the numbers in the equations into the matrix. Put the coefficients of the variables of the first equation in Row 1 in the matrix. Place the coefficients of the variables of the second equation in Row 2 of the matrix. Place a vertical line after the coefficients. Put the constants (the c-terms) in the equations in the third column of the matrix.

$$\begin{bmatrix} 2 & 5 & | & 0 \\ 1 & 1 & | & -3 \end{bmatrix}$$

Step 3: Put the numbers in the matrix to the left of the vertical line in Reduced Row Echelon Form.

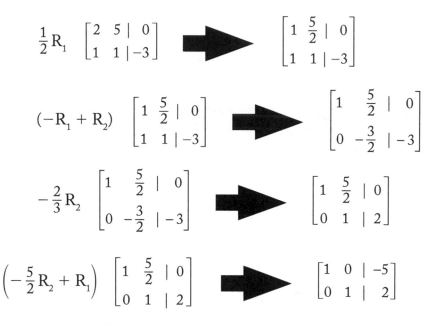

$\frac{1}{2}R_1$ $\begin{bmatrix} 2 & 5 & | & 0 \\ 1 & 1 & | & -3 \end{bmatrix}$ ➡ $\begin{bmatrix} 1 & \frac{5}{2} & | & 0 \\ 1 & 1 & | & -3 \end{bmatrix}$

$(-R_1 + R_2)$ $\begin{bmatrix} 1 & \frac{5}{2} & | & 0 \\ 1 & 1 & | & -3 \end{bmatrix}$ ➡ $\begin{bmatrix} 1 & \frac{5}{2} & | & 0 \\ 0 & -\frac{3}{2} & | & -3 \end{bmatrix}$

$-\frac{2}{3}R_2$ $\begin{bmatrix} 1 & \frac{5}{2} & | & 0 \\ 0 & -\frac{3}{2} & | & -3 \end{bmatrix}$ ➡ $\begin{bmatrix} 1 & \frac{5}{2} & | & 0 \\ 0 & 1 & | & 2 \end{bmatrix}$

$\left(-\frac{5}{2}R_2 + R_1\right)$ $\begin{bmatrix} 1 & \frac{5}{2} & | & 0 \\ 0 & 1 & | & 2 \end{bmatrix}$ ➡ $\begin{bmatrix} 1 & 0 & | & -5 \\ 0 & 1 & | & 2 \end{bmatrix}$

Step 4: Place the elements of the matrix back into the linear equations and solve.

$$1x + 0y = -5$$
$$0x + 1y = 2$$

Now solve these two equations.

$$x = -5$$
$$y = 2$$

BRAIN TICKLERS   Set # 70

Solve the following systems of equations using matrices.

1. $x + 2y = 9$ and $4x + y = 8$

2. $2x + 10y + 4 = 0$ and $-10x + 2y - 20 = 0$

3. $x + 2y = 7$ and $-x + 6y = -3$

(Answers are on page 283.)

Functions

What's a function? A function is something that operates on one number to create another number. Think of a function as a magic box; one number enters the box and another number exits it.

Input   Function   Output

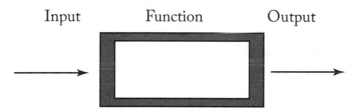

The number entering the box is called the input. The number exiting the box is called the output. What happens inside the box is the function.

Look at this chart. The numbers 0, 1, 2, 4, and 7 were put in a box. The numbers 3, 4, 5, 7, and 10 came out of the box. What do you think is happening inside the box?

Input	What's the Operation	Output
0		3
1		4
2		5
4		7
7		10

Look at the first number in the input column. When zero is the input, three is the output. $0 + 3 = 3$

Is $+3$ the correct operation?

Check the next input number. $1 + 3 = 4$

Is the magic box adding three to each number that enters the box?

Add three to each of the input numbers to see if you get the output number.

Input	What's the Operation	Output
0	+3	3
1	+3	4
2	+3	5
4	+3	7
7	+3	10

Yes. To write this as a function, write $f(x) = x + 3$.

(1+2=3) MATH TALK!

$$f(x) = x + 3$$

Read this as "f *of* x *is equal to* x + 3"

f *is the name of this function.*

You can also name a function g, h, or any other letter.

f *of* x *tells you the function operates on the letter* x.

$f(x) = x + 3$ *tells us the function takes* x *and adds 3 to it.*

Now evaluate this magic box. What is the operation in the box?

Input   Function   Output

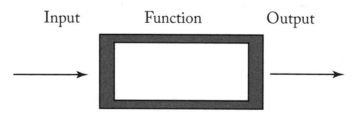

Input	What's the Operation?	Output
2		4
4		8
8		16
10		20

Look at the first number in the input column. What operation would change 2 to 4? You could add two to two to get four. If the magic box added 2 to all the input numbers, then the output numbers would be 4, 6, 10, and 12. These numbers do not match the output numbers, so the operation is not $x + 2$. Look more carefully at the input and output numbers. The output numbers are twice the input numbers. The magic box is doubling each number.

The function is $f(x) = 2x$. Read this as f *of* x *is equal to* 2x.

Sets and functions

A function operates on a set. A set is a collection of things. It could be a collection of numbers, or dogs, or letters. The things in a set are called members or elements.

When a function operates on a set, it could operate on a finite set, that is, a set with a limited number of elements, such as the set $\{1, 2, 3, 4\}$. Or a function can operate on an infinite set of numbers such as $\{1, 2, 3, \ldots\}$. What makes a function special is that it must work for each number you put in the box. It must also yield only one output for each input. In other words, you can't put one number in the box and have two numbers come out.

What does that mean?

If $f(x) = x^2 + 1$ is evaluated for the set of numbers $\{0, 1, 2, 3\}$, the result is the set of ordered pairs, $\{(0, 1), (1, 2), (2, 5), (3, 10)\}$. The function or operation works for each and every value of x and each and every value of x has a unique result.

BRAIN TICKLERS   Set # 71

Which of the following sets of ordered pairs are functions?

1.   {(1, 2), (3, 4), (5, 6), (7, 8), (9, 0)}

2.   {(1, 0), (2, 0), (3, 0), (4, 0), (5, 0)}

3.   {(0, 2), (0, 4), (0, 6), (0, 8)}

(Answers are on page 283.)

Domain and range

What goes into the magic box is the input. All the numbers that compose the input are called the *domain of the function*. The *x*-values are the *domain* of the function.

What comes out of the magic box is the output. All the elements that compose the output are called the *range*. The $f(x)$ values are the range of the function.

The elements in the domain are the independent variables. The elements in the range are the dependent variables. The elements in the range depend on the elements of the domain.

VISUALIZE IT

$f(x) = 2x$ **for the set of numbers**

{1, 2, 3, 4} is {(1, 2), (2, 4), (3, 6), (4, 8)}.

$x = \{1, 2, 3, 4\}$	$f(x) = \{2, 4, 6, 8\}$
Input	Output
Domain	Range
Independent Variable	Dependent Variable

Evaluating a function

To evaluate a function, substitute each of the *x*-values into the function. The result will be $f(x)$.

   If $f(x) = x - 1$, what is $f(4)$?

This function is being evaluated for only one number.

Just substitute 4 for x.

$f(4) = 3$

The answer is the ordered pair $(4, 3)$.

Watch as this function is evaluated.

$$f(x) = x - 2 \text{ for } x = \{1, 2, 3, 4, 5\}$$

Make a chart and fill in the x-values.

Next fill in the $f(x)$ values. Just substitute the value for x in the expression $x - 2$.

x	$f(x)$
1	-1
2	0
3	1
4	2
5	3

The function is the set of ordered pairs,

$$\{(1, -1), (2, 0), (3, 1), (4, 2), (5, 3)\}.$$

Watch as this function is evaluated.

$$f(x) = x^2 - 4 \text{ for the set } \{-2, -1, 0, 1, 2\}$$

Substitute each value of x in the expression $x^2 - 4$ to find $f(x)$.

x	$f(x)$
-2	0
-1	-3
0	-4
1	-3
2	0

The function is

$$\{(-2, 0), (-1, -3), (0, -4), (1, -3), (2, 0)\}.$$

The result of applying a function to a set is a set of ordered pairs.

BRAIN TICKLERS   Set # 72

Evaluate the following functions.

1. $f(x) = \dfrac{x}{2}$ for $x = \{-2, 0, 2, 4, 6, 8, 10\}$

2. $f(x) = -3x + 1$ for $x = \{1, 10, 100\}$

3. $f(x) = x^2 + 100$ for $x = \{-10, -5, 0, 5, 10\}$

(Answers are on page 283.)

Vertical line test

If you graph an equation or you are shown the graph of an equation, you can determine whether the graph is a function by what's called the Vertical Line Test. Look at the graph and see if there is anywhere that you can draw a vertical line that intersects the graph in more than one place. If you can, it is not a function.

Example:

Look at this graph of a parabola. Can you draw a vertical line that will intersect this parabola in more than one place?

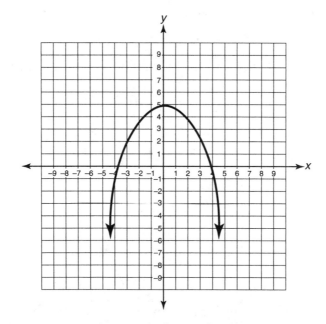

No, you cannot, so this parabola is a function.

Example:

Look at this graph of a square. Can you draw a vertical line that will intersect this graph in more than one place?

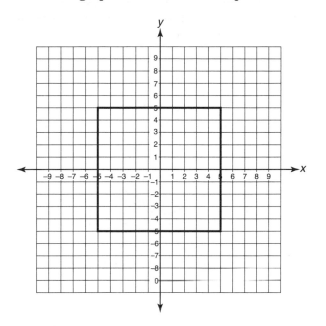

Yes, you can. Any vertical line that intersects the square will intersect it in at least two points. This graph of a square is not a function.

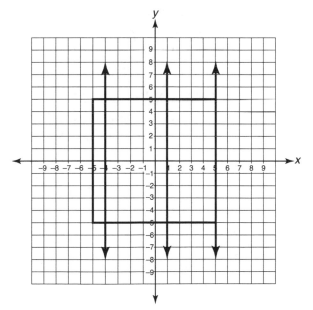

BRAIN TICKLERS   Set # 73

Look at the following graphs. Use the vertical line test to determine whether they are functions. Answer Yes or No.

1.

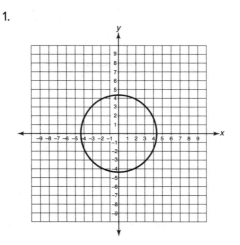

2.

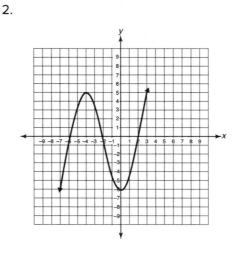

3.

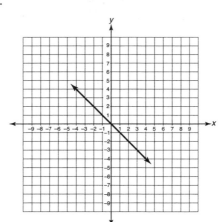

4.

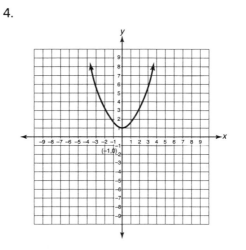

(Answers are on page 283.)

BRAIN TICKLERS—THE ANSWERS

Set # 64, page 253

1. 6
2. 24
3. 60
4. -10
5. -20
6. -80
7. $\frac{1}{2}$
8. $\frac{11}{2}$
9. $\frac{27}{2}$

Set # 65, page 257

1. 2
2. 24
3. 192
4. -1
5. 5
6. -5
7. $\frac{1}{2}$
8. 16
9. $\frac{1}{2}$

Set # 66, page 259

1. 4
2. 21
3. 1
4. 35 and 35
5. 20
6. 1 10 45 120 210 252 210 120 45 10 1
7. 32 or 2^5
8. 2^{49}

Set # 67, page 263

1. 9
2. 1,001
3. a^5

4. $(a + b)^4 = a^4 + 4a^3b + 6a^2b^2 + 4ab^3 + b^4$

5. $(a + b)^9 = a^9 + 9a^8b + 36a^7b^2 + 84a^6b^3 + 126a^5b^4 + 126a^4b^5 + 84a^3b^6 + 36a^2b^7 + 9ab^8 + b^9$

6. $(a + b)^{11} = a^{11} + 11a^{10}b + 55a^9b^2 + 165a^8b^3 + 330a^7b^4 + 462a^6b^5 + 462a^5b^6 + 330a^4b^7 + 165a^3b^8 + 55a^2b^9 + 11a^1b^{10} + b^{11}$

Set # 68, page 265

1. 4×2
3. 2nd
5. 0.5

2. 2nd
4. 3

Set # 69, page 269

1. $\begin{bmatrix} 1 & 0 \\ 0 & 1 \end{bmatrix}$
2. $\begin{bmatrix} 1 & 0 \\ 0 & 1 \end{bmatrix}$
3. $\begin{bmatrix} 1 & 0 \\ 0 & 1 \end{bmatrix}$

Set # 70, page 273

1. $x = 1, y = 4$
2. $x = -2, y = 0$

3. $x = 6, y = \dfrac{1}{2}$

Set # 71, page 276

1. Yes
2. Yes
3. No

Set # 72, page 278

1. $\{(-2, -1), (0, 0), (2, 1), (4, 2), (6, 3), (8, 4), (10, 5)\}$

2. $\{(1, -2), (10, -29), (100, -299)\}$

3. $\{(-10, 200), (-5, 125), (0, 100), (5, 125), (10, 200)\}$

Set # 73, page 280

1. No
2. Yes
3. Yes
4. Yes

Index